THE BLACK DEBACLE

From a Thundering Voice to a Confused Whimper

(Recollections and Observations on the United Mine Workers and Collective Bargaining in the Coal Industry)

BY MELVIN TRIOLO

mpc

McClain Printing Company
Parsons, West Virginia
1991

International Standard Book Number 0-9628040-0-2
Library of Congress Catalog Card Number 90-91419
Printed in the United States of America

Contents

Preface

This book is dedicated to my late father and mother who came to this country nearly a century ago in pursuit of the American dream, settled in the coalfields of West Virginia and raised their family with love and by example to be responsible and contributing American citizens.

It recounts some intimate details in the early struggles, the ascendancy, and the decline of the United Mine Workers Union as the premier labor organization of this country. It is written from my own vantage point of over seventy-five years of experience in the coalfields—from my birth to the present; and as an involved member and official of the union, and as a management representative in collective bargaining.

It highlights the life, leadership and accomplishments—and the foibles and warts, too—of the greatest labor leader of all time, John L. Lewis, the Thundering Voice.

Much of what is recounted in this book comes from my own recollections and conversations with key figures. I have also quoted extensively from newspaper accounts and other published materials that were pertinent to and descriptive of events as they occurred.

Any opinions or views expressed in this book are mine, and I take full responsibility, realizing that there may be some who may on occasion differ.

It is my fervent hope that what I have written will not only contribute to the love of the coal industry but will also be of some help in bringing forth an understanding of its problems and spur leadership in surmounting them.

CHAPTER I

The Vantage Point

My grandfather, on my father's side, was Carmelo Triolo. His father and mother immigrated from the Island of Corsica and settled in Motta San Giovanni at the turn of the eighteenth century about the time another Corsican, Napoleon Bonaparte, had left Corsica and was making history in France and Europe. To this marriage were born five sons, and they were all raised in Reggio Calabria, Italy.

My great-grandfather was also named Carmelo Triolo. After his sons were grown, four of them departed southern Italy to find a future of their own. After the death of my great-grandfather and great-grandmother, my grandfather was unable to locate any of his brothers and thereafter never knew their whereabouts.

My grandfather, Carmelo, married Angela Verduci. Out of this marriage, four children were born, two daughters, Nina and Angela, and two sons, Antonio and my father, Giovanni. My father was born on November 22, 1879, at Motta San Giovanni, Reggio Calabria, Italy. The family was raised on a hillside farm that my grandfather had inherited from my great-grandfather.

My father and his brother and sisters had very limited educations; however, they all learned to read and write.

Around 1899, my uncle Antonio Triolo and some of his boyhood friends left together for America. Their destination was Kimball in McDowell County in southern West Virginia. There was quite a large group of young Italian men from my father's hometown who worked in the mines around Kimball, but particularly in the town next to Kimball called Vivian.

Antonio was employed with a coal company at Vivian that mined the Pocahontas #3 seam of coal. This was later known as the best seam of coal being mined in North America.

In 1900, at the age of twenty-one, my father and a distant

relative decided to leave their homeland and come to America. They read that America was a fast growing country with a world of opportunity. My father had received money from Antonio to come to America, and my cousin received money from a relative of his who was working in the mines at the time in Vivian, West Virginia. So, they left home and made their way to Naples, Italy's largest seaport. They had very little money and only a change of clothes in their suitcases. Finally, they were able to secure transportation on a boat going to America, and they both agreed to work on the boat doing odd jobs to pay for their transportation. They arrived at Ellis Island in New York in the latter part of 1900.

The trip to cross the Atlantic from Naples to New York took twenty-eight days and nights. Since neither my father nor his cousin could speak any English, they finally became acquainted with an Italian who was working at Ellis Island, and they inquired of him where they could find work. He suggested they go to Buffalo, New York. It was a new town where there were many steel mills being built, and there was all kinds of work. They got on a train and arrived in Buffalo that evening. The next day, they found employment with the city of Buffalo, laying streetcar lines for twenty-five cents a day. Since they had no skills, they had to take whatever unskilled work they could get.

On September 6, 1901, the city of Buffalo opened the Pan-American Exposition. This brought about a large celebration because President McKinley came to Buffalo on this date to address the exposition. As President McKinley was getting off the train, he was shot by a man who extended his hand as if to shake hands with him. McKinley was rushed to a hospital and died eight days later on September 14, 1901. My father and his cousin were among the thousands who were present and witnessed the assassination.

The two young men continued to work in Buffalo until 1902 when my father received a telegram that his brother, Antonio, had been seriously injured in a mine accident at Vivian. They both quit their jobs and caught a train that took them from Buffalo, New York, to Kimball, West Virginia.

Upon arrival in Kimball, my father was told by some of his brother's friends that his brother had died and was buried in the Kimball Cemetery. After spending about a month in the vicinity of Kimball, the two young men learned of the new coalfields being opened in Logan County, West Virginia. They left Kimball and rode the N&W Railroad train to Dingess, West

Virginia, which was as close as they could get to Logan. There were no facilities for transportation from Dingess to Logan. As a result, they walked all day and arrived in Logan about 4:30 in the afternoon. Neither one could speak nor understand the English language; however, they managed to get a room at a rooming house for the night. The next day, they went to the site where the Chesapeake and Ohio Railway was building a railroad station for the town of Logan. They were hired by the railroad company and were paid two dollars per day for an eight-hour day. They continued their employment with the railroad company after they finished the railroad station, which is the present-day location of the city hall for the city of Logan. The railroad spread its crews out and built a railroad bridge across the Guyandotte River and continued to lay their tracks toward Monitor Junction. When they got to Monitor Junction, the railroad company split the crew and took part of the crew to continue laying tracks toward Holden; another part of the crew was taken to lay tracks toward Island Creek.

It turned out that my father and cousin were in the crew that continued to lay the railroad tracks toward Island Creek. When the railroad tracks reached near Wilkinson, a Robinson family from Shamokin, Pennsylvania, opened up three mines. Mine #1, which they called the Beechwood Mine, and Mine #2 opened up the Yuma side of Island Creek. The #3 Mine was located at Wilkinson. The Robinson's named their company the Monitor Coal and Coke Company. This was about 1906.

My father and cousin were first employed in the coal mines at the #2 Yuma Mine. They continued their employment with Monitor Coal and Coke Company until 1912, at which time they both returned to Italy for a visit with their families. While there, my cousin married. My father became engaged to my mother, Maria Azzara, who was from the same small town where my father and cousin stayed. After being gone six weeks, the two men returned to Logan, and continued their employment with Monitor Coal and Coke Company.

In the early part of 1913, both my father and cousin sent money for his cousin's wife and son, my then-to-be-mother and another cousin who also had a son, to come to America. They, too, arrived at Ellis Island and came to Logan on a train from New York. My mother and father were married the next day at the courthouse in Logan by the Reverend J. Green McNeely, who served as the county clerk.

In the meantime, each family had rented a company-owned house in Yuma. Later that year, my father and cousin were transferred to the #3 Mine at Wilkinson, and my mother and father moved into one of the new houses that the company had built for the employees of the #3 Mine.

On January 14, 1914, my mother gave birth to their eldest son, Melvin. Later came a second son, Antonio (Tony), a third son, Giovanni (John) and a fourth son, Giuseppe (Joe). All four sons were born in the same house that my mother and father lived in at Wilkinson.

All four of the sons were delivered by the company doctor, Dr. Sidney B. Lawson of Logan. Dr. Lawson wrote a book titled "Fifty Years as a Country Doctor." He later entered politics and was a county court member. He was highly respected and served for many years on the county court. He was part owner of the first hospital in Logan, known as the Hatfield-Lawson hospital. He was also a member of the West Virginia House of Delegates, and he introduced a resolution into the House of Delegates in 1895, creating Mingo County from part of Logan County. Dr. Lawson retired in his late seventies. He sold the stately mansion that he had built and lived in on Cole Street in Logan and retired to Florida. He and his wife both died in Florida and were brought back to Logan; their bodies were interred in the Logan Memorial Park.

About 1919 or 1920, my father's cousin moved his family to Monaville, where a new operation by Island Creek Coal Company was being started. He went to work at the #11 Island Creek Mine and later talked my father into coming to work there.

In the latter part of 1921, my father moved our family to Monaville into a house next to my cousins where my mother gave birth to my three sisters, Angeline, Catherine and Helen. Angeline is the oldest, Catherine was next and Helen is the youngest. All three of my sisters were born in the same house that we lived in at Monaville. Dr. Frederick Round delivered my sister Angeline, and Dr. John McNabb delivered my sisters Catherine and Helen.

During the late 1920s, we all attended the elementary school at Monaville and later attended what is now Logan Central Jr. High School. In the late 1920s and 1930s, it was known as Logan Jr. High School. I graduated from Logan High School in 1934.

After finishing high school, I attended Marshall University and Morris Harvey College in Charleston, West Virginia, now known as the University of Charleston, majoring in Political Science with a minor in Criminology. Starting at age sixteen during summer school recesses, and later after finishing school—I went to work in the mines as a coal loader. I worked underground with my father as his "buddy." I had to go into the mines to help support our large family. This was at a time when the country was just beginning to come out of the "Big Depression." My father was barely making an existence for his family, and the mines were only working one or two days a week.

Thereafter, the next fifty-five years of my life have been exclusively devoted to the coal industry—with the United Mine Workers Union; with coal management in industrial relations responsibilities; in community affairs and in politics, as will be set forth in the pages that follow.

I cannot stress too strongly my admiration and love for my parents. Both of them left their home and families in southern Italy and immigrated to this country. In 1901, it took tremendous courage for my father and his cousin to seek employment after arriving in Logan. You can just imagine two immigrants, who could not speak the English language, who walked from Dingess to Logan to seek employment. It took a lot of guts for them to stay here under those circumstances and make a living at $2 per day for an eight-hour day. My father returned to the place of his birth in 1913 and again in July of 1950. These were the only two times that my father went back, and these were short visits. His home was in the United States, and he exemplified the American dream—as an honest, hardworking and freedom loving coal miner.

My father died on July 6, 1962. He was eighty-three years old, and he had worked in the mines until he was seventy-three years old. All of his life he worked for only two coal companies—Monitor Coal and Coke Company and Island Creek Coal Company.

My mother arrived here in 1913 after leaving her home, two sisters, a brother, and her mother and father. She never returned to her original homeplace; therefore, she never saw her mother and father, her brother or sisters after she left Italy. She died in Cabell County on December 30, 1965.

Both my mother and father worked very hard throughout their lives to raise a family of seven children; four sons and three daughters.

We had a hard life. My mother kept us together, and without her love and patience, it would have been very hard for us to succeed in life. All of the successes that I have enjoyed during my lifetime are due to the loving care I received from my mother and father. The discipline and training they provided for me and my brothers and sisters is responsible for whatever successes we have achieved, and I dedicate this book in admiration and love to their memory.

CHAPTER II

Career in Coal

After finishing school, I was again employed at Island Creek Coal Company as a coal loader. Although it was hard work, I enjoyed working in the mines. Most of my mining career was with Island Creek Coal Company. In June of 1936, I was laid off due to bad marketing conditions and lack of demand for coal.

In July of 1936, I was employed by the West Virginia Coal and Coke Corporation at its Rossmore Mine as track helper. I continued to work in this capacity until January of 1937. Due to the union checkweighman resigning his position, I ran for the position of checkweighman and was elected. In this position, my responsibility was to see that the coal loaders, and machine and drill men received their proper weight in the mining and cutting of coal. This was one of the best jobs I ever had.

In the latter part of September in 1937, I resigned my position as checkweighman to accept a position as a field representative in the Logan office of the United Mine Workers of America, District 17, which was headquartered in Charleston, West Virginia. I continued in this capacity until August 1, 1950. Then, I was elevated to the position of chief representative of the Logan field, succeeding the late Millard Cassidy, who died July 1, 1950. I continued as the chief representative until my resignation on October 5, 1955.

Prior to my resignation, I was called to Washington by President John L. Lewis for an audience with him at his request. At this meeting in January, he complimented me on my performance as chief representative in the Logan field. He was highly complimentary of the method by which I was handling the collection for the Welfare Fund—there were no companies in the Logan field delinquent in their collection of this money. He also complimented me on the method by which we handled our grievances. He told me that he was going to retire William

Blizzard, president of District 17, on April 1, 1955, and that he was going to place me in charge of the district on this day. When asked how well I knew Governor Marland, I responded by saying that I knew him quite well. Mr. Lewis proceeded to tell me that upon my return to Logan, he wanted me to handle the legislative business of the organization in West Virginia; since the West Virginia Legislature was in session, he wanted me to work closely with Governor Marland in completing his program. He called Governor Marland and told him that I was in his office, and he asked the governor if he knew me. The governor responded by saying that he knew me quite well. Mr. Lewis proceeded to tell him that I was coming to Charleston the next morning, and he wanted to know if he would give me an appointment to see him before I proceeded on to Logan. Governor Marland responded by asking Mr. Lewis what time I would be arriving at the airport, and Mr. Lewis told him about 11:15 a.m. Mr. Lewis also told the governor to contact me immediately if anything came up in the legislature that he thought the mine workers would be interested in—that I would be handling the affairs of the union in District 17.

Governor Marland told Mr. Lewis that he would have his valet, Mr. Ferguson, pick me up at the airport and take me to the Governor's mansion where we would have lunch together. Mr. Lewis told the governor that he appreciated it and that he would have me wait at the airport for Mr. Ferguson. Upon my arrival in Charleston, the governor's valet picked me up, and we proceeded to the mansion. The governor was waiting for me, and we had lunch. During our conversation, he said he was surprised to learn about Mr. Lewis's action concerning Bill Blizzard. He asked me if I knew what had happened between Blizzard and Mr. Lewis. I told him I thought it was over Raymond Lewis's and William Blizzard's fight in the district office that had taken place a couple of weeks before I went to Washington. I told him that it was my understanding that the fight was over Blizzard's personal mail being opened in Raymond Lewis's office. There in his office, Raymond Lewis had called Bill Blizzard an "S.O.B.," and Blizzard had hit him between the eyes, blacking both his eyes and beating him pretty severely. I had also been told that three of the Kanawha fieldmen were in the office, and not one of them attempted to separate the two men.

About April 1, I received a letter that was addressed to all of the representatives of District 17 announcing that Raymond O. Lewis had been designated as president of District 17, effective May 1, 1955. This was altogether different from Mr. Lewis's conversation with me in his office in January of 1955.

Shortly after June 1, I was ordered to Charleston by Raymond Lewis. When I arrived at his office, present with him were Charlie Kiser, who was in charge of the Williamson office of the UMWA and Charlie Neal, who had just been named vice-president of District 17 at the same time that Raymond Lewis had been named president. Raymond Lewis proceeded to tell me that effective the first of the month, I would report to Mr. Kiser in the Williamson office and that Mr. Kiser would assign my duties and be in charge of both the Logan and Williamson offices. Raymond Lewis asked me if I was going to accept this assignment. I responded that I would be there on the first of the month and would report to the Williamson office. This seemed to surprise him. He also told me that effective the first of the month, my salary would be reduced by one hundred dollars, and he wanted to know how I felt about this. I told him that as long as I worked for the union, I would carry out their instructions, and I would be in Williamson the first of the month. I had made up my mind that I was going to seek another audience with John L. Lewis to find out what had happened after I left his office in Washington particularly after having received his glowing compliments.

I reported to Williamson, and Kiser assigned me to organize the Wolf Creek Collieries mine, a nonunion operation located on Wolf Creek that employed about ninety men. I went about doing what work I could, and I later found out that another nonunion mine was located on Crum Mountain that worked about ninety men in what was known as the Saunders Coal Company. Still later, I found in the course of my work more nonunion operations in Kermit on the Kentucky side where there were three nonunion mines. In the meantime, I wired Mr. Lewis seeking another audience because I wanted to find out from him what had happened to me. I did not receive any response to my telegram until a month later, when I received a letter from Mr. Lewis requesting that I write a letter of complaint and mail it to him. I responded to Mr. Lewis's letter. However, I pointed out to Mr. Lewis that the nature of my business was such that I could not and would not put it in a letter, and that during my eighteen years with the union, I had never requested an audience with him. I had, in times past, had several audiences with him which were all at his request, and in all fairness to me and based on my tenure with the United Mine

Workers, I thought that I at least deserved an audience. He responded about two weeks later and stated that the only way he would grant me an audience was for me to get approval from Raymond O. Lewis, president of District 17, that I be granted an audience with Mr. Lewis. After receiving his letter, I went to Charleston and met with Raymond Lewis. Raymond wanted to know the nature of the complaint that I wished to discuss with John Lewis. I told him that I wanted to discuss with John Lewis the conversation I had had with him when I was in his office in January, at which time I was there at his request. After discussing this matter with R. O. Lewis, he told me that he would write John Lewis a letter that afternoon recommending that he give me an audience and that he would send me a copy of the letter.

The next afternoon upon my return from Williamson, I went by the post office and received a copy of Mr. R. O. Lewis's letter to John Lewis. The substance of the letter read that he and I had discussed my complaint and that he saw no necessity for John Lewis to give me an audience. I wired John L. Lewis again and told him that R. O. Lewis had lied to me because he had said he would recommend that he give me an audience and he did not recommend this in his letter. I, again, reiterated my service with the union and stated that common courtesy should allow me the privilege of another audience with him. Mr. Lewis wired back and told me to be in Washington on September 13, 1955. I met with him in his office, and he wanted to know what my problems were. I told him that I first wanted to find out what had happened that had changed his mind about my succeeding William Blizzard. He told me that it was his prerogative that, whomever he wanted to name as president, he had the complete authority to do so. I also brought to his attention that he had told me what an outstanding job I was doing as the chief representative in the Logan Field and that Raymond Lewis had removed me as the chief representative and sent me to Williamson. In addition, I told him I had never received a wage cut in my life until I went to work for the United Mine Workers. He wanted to know who had cut my wages, and I told him that Raymond Lewis had.

On October 5, 1955, I resigned as an officer of District 17. On October 7, 1955, I was called into Raymond Lewis's office, and at that time, I demanded that he pay me in full through October 5, 1955. He had Mrs. Waldo, his secretary, direct the bookkeeping department to write the check and give it to me. I left his office and came back to Logan. It was so close to

Christmas that I did not then seek other employment. About the first or second of January, I received a check in the mail from District 17 paying me from October 5 to December 31, 1955. I am assuming that the reason I was paid through the end of the year was probably because John L. Lewis had made inquiry about the reduction of my salary when I was transferred to Williamson.

On January 15, 1956, I accepted employment as the Director of Industrial Relations and Personnel for Jewell Ridge Coal Corporation in Tazewell, Virginia. I continued this employment until June 1, 1962. I left employment with Jewell Ridge to accept employment as the secretary-treasurer of the Logan Coal Operators Association in Logan, West Virginia. I continued my employment with the coal association from June 1, 1962, through September 1, 1985, when the coal association was dissolved.

CHAPTER III

John L. Lewis—
The Thundering Voice of Labor

About 1870, the coal miners in Wales began to immigrate from their homeland to North America, seeking work in the coal mines. Among some of the immigrants were the Watkins family and Thomas Lewis.

The Watkins family and Thomas Lewis settled in the small mining town of Lucas, Iowa, and, shortly after their arrival in Lucas, Iowa, Thomas Lewis married Louisa Watkins. Born out of this marriage were six sons and two daughters. The eldest was John Llewellyn Lewis, born February 12, 1880, and the others in the following order: Thomas, Jr., A. D. (Denny), Howard, Hattie, George, Sarah and Raymond.

"When young John was two years old, his father helped to lead his fellow miners in a bitter strike against their employers, the White Breast Fuel Company. He fought the company so bitterly that it became apparent that, regardless of the outcome, he would never again be employed by it. Although the strike was won, Thomas Lewis stood alone, watching his friends and fellow miners return to work while he, who had fought so unreservedly for them, was forgotten. But the White Breast Fuel Company did not forget, and Thomas Lewis was placed on their blacklist as a radical and unionist.

"From that time on Thomas Lewis was condemned to a wandering exile, seeking here and there for work to support his family. Now there was another child, Thomas, Jr. and the Lewises moved to Colfax, Iowa, where for a time Tom Lewis worked as a night watchman, then again as a miner in an adjacent mining town until the blacklist reached his employers. Then the Lewis family moved to the state capital, Des Moines. Thomas Lewis again began to work as a night watchman and later was appointed custodian of the city jail. Denny Lewis was born in 1889. Three years later came Howard Lewis and, in 1894, Hattie Lewis. Eventually two more sons and one more daughter was born to Louisa and Thomas Lewis, so that the Lewis family consisted of six sons and two daughters.

"Young John was an aggressive, pugnacious youngster. John first attended the Grant Park School, which was close to the Iowa State fairgrounds. From there he went to the Lincoln School in East Des Moines, then the old Washington School and finally attended briefly a junior high school close to the state capital. He rebelled against the formal curriculum of school and quit before he finished the eighth grade.

"In 1897 the blacklists were destroyed and Thomas Lewis was free to return to the coal fields. At once the Lewis family left Des Moines for their home town of Lucas, Iowa, where John Lewis, now seventeen years old, his brother Thomas, and his father all began to work in the Big Hill Coal and Mining Company. Ventilation of the mines was unheard of in those days. They sweated it out, working ten to eleven hours a day, working in the suffocation of power fumes.

"But still the Lewis family was much happier than they had been in Des Moines. Here they were at home among their own folk, Welsh miners who were hardy, rugged people. They were a unique breed, dramatic both in speech and gesture, a strange paradox of sentimental romanticism and hardened realism. They could be tender and yet ruthless. They sang their old Welsh folk songs and quoted Shakespeare in their daily conversations. To understand them was to understand much of what went into the makeup of John L. Lewis.

"His energy was boundless. After the day's work was done, John organized and managed both a debating team and a baseball team. He was a human sponge soaking up both information and knowledge from his avid reading. Lewis loved to read and devoured book after book, practically all of them of the sensational stripe. His interests, like his reading habits, were very disorganized until he went to a party and became attracted to young Myrta Edith Bell. (Note: It has always been assumed that Lewis met Myrta Bell at this party. However, Lewis told the writer, "I had already known Mrs. Lewis (Myrta Bell) for fifteen years before I married her.")

"Myrta Bell was the eldest daughter of an Ohio doctor who had moved to Lucas when she was ten years old. The Bell family tree had its roots in the American colonies before the American revolution. It was peopled primarily by scholars, educators, and physicians. It possessed stability and security, and these were reflected in Myrta Bell. She was a calm girl, possessed of an inner security that bloomed into serenity as the years passed. Few ever realized her incredible inner strength. She was in a profound sense stronger than John Lewis. Underneath the tender gentility was a will of inflexible steel.

"Myrta Bell organized Lewis's reading habits and introduced him to Dickens and to Homer and other classics. She encouraged his interests, guided him, and loved him. Myrta Bell was to become the most important single force in the life of John L. Lewis."[1]

When John Lewis started thinking about the best starting point for his career while living in Lucas, Iowa, he decided that he would be able to accomplish more by moving someplace where he could get better exposure than he could get in Lucas, Iowa, and in 1909, John and Myrta Lewis migrated to Panama, Illinois.

"The Lewis family followed them very shortly. Here in the heart of the Montgomery coal fields, Lewis began passionately to orate on the injustices besetting the miners. The five Lewis brothers formed a loyal clique and vigorously led the applause. Lewis's machine of his five brothers soon captured control of the union local, and John was elected president within a year.

"With his wife's invaluable aid, Lewis began to write and practice the art of speechmaking that would both arouse his listeners and establish his dominance as a leader. Many of his gestures, mannerisms, vocal inflections, and phrases were practiced in the privacy of his home and on an audience of one, Myrta Lewis. His wife would patiently correct, modify, change, and polish his delivery. Then as all intelligent wives have done and always will do, she applauded her husband for *his* finished delivery.

"Then came the Cherry disaster, a mine explosion and fire that killed one hundred and sixty miners. Lewis stood before the state legislature like a mad bull. He raged and banged his fist on the rostrum. He did everything but personally promise to physically demolish each legislator. Intimidated, they passed bill after bill, including a workmen's compensation law. News of this victory traveled far; and even Samuel Gompers, founder and president of the AF of L, asked about this young man in Illinois."[2]

1. Saul D. Alinsky, *John L. Lewis: An Unauthorized Biography*, pp. 15-17.
2. Saul D. Alinsky, *John L. Lewis: An Unauthorized Biography*, pp. 21-22.

When John Lewis got to Panama, Illinois, he went to work in the mines around Panama. Due to his activities in the Mine Workers Union, he was made the union's legislative representative in the Illinois Legislature, and he handled this job so effectively that Samuel Gompers, President of the American Federation of Labor (AFL) in 1911, appointed him as a field representative and legislative agent for the AFL. He served in this capacity until 1916 when he was employed as International Statistician of the United Mine Workers (UMW) by John P. White, the International President.

Later that same year, John White was appointed as an arbitrator to handle labor disputes and resigned as president of the United Mine Workers of America (UMWA). Frank J. Hayes, John White's vice-president, succeeded to the presidency, and he appointed John L. Lewis as International Vice-President.

John L. Lewis continued to serve in this capacity until Frank J. Hayes resigned because of health reasons. In 1919, John Lewis presided over the 1919 convention of the UMWA in Indianapolis, Indiana. In 1920, he was elected to a five-year term as president of the UMWA and was thereafter re-elected for eight successive five-year terms until he retired in 1960. Thomas A. Kennedy, International Secretary-Treasurer, succeeded John Lewis, and John Owens, president of District Six in Ohio, succeeded Mr. Kennedy as secretary-treasurer of the UMWA.

In John L. Lewis's campaign in 1920, he was bitterly fought by an element in the Mine Workers Union that was partly Socialist and partly members of the Communist party. After assuming the office of president, Lewis removed some of his most bitter enemies from office, particularly John Brophy, who was president of District Two in Pennsylvania and who had run against John Lewis for the presidency. Brophy was an out-and-out Socialist, as were Alexander Howat, president of District Fourteen of Pittsburgh, Kansas; Adolph Germer, an officer of the union in Illinois, District Twelve; Frank Farrington, president of District Twelve; and many others too numerous to identify.

John L. Lewis, in 1924, had the constitutional convention amend the union constitution to prohibit members of the Communist party, members of the Industrial Workers of the World (IWW) and the National Organization of the Woodsmen of the World, from holding membership in the UMW Union. This was the first time that a provision outlawing these subversive organizations was written into the constitution of the UMW, and it has been in every constitution since that time.

After Mr. Lewis had the constitution amended to prohibit membership in these organizations, he kicked out of the UMW Union several highly intelligent persons who belonged to some of these organizations. In addition to those already mentioned, there were Alan Haywood, Powers Hapgood, and several others who were members of the Socialist party and advocated principles of the Socialist party in the UMW Union.

After Mr. Lewis gained firm control of the mine workers organization, in 1924, he was faced with the expiration of the UMW contract with the Central Competitive Field, which comprised coal mines in Illinois, Indiana, Pennsylvania, Ohio and northern West Virginia.

At the 1924 convention of the mine workers which was held in Indianapolis, Indiana, the Scale Committee of the union was authorized to meet with the bituminous coal operators in Jacksonville, Florida, to try to negotiate a four-year wage agreement. On February 18, 1924, the bituminous operators and the UMW approved a three-year wage agreement known as the Jacksonville Agreement, so-called because it had been negotiated in Jacksonville, Florida. "It was looked upon as a victory for the union since the operators had insisted that they would never sign a contract binding them to high wages for more than one year. This agreement was to extend to April 1, 1927."[3]

This agreement provided the highest wage scale in the Central Competitive Field between the operators and the mine workers. It was brought about by the Secretary of Commerce, Herbert Hoover, who put the pressure on the coal operators in the Central Competitive Field to sign the agreement. In return for Secretary Hoover's help in getting the agreement signed, John L. Lewis pledged his and the mine workers support to Mr. Hoover in his 1928 candidacy for the presidency of the United States. However, by the time Hoover was elected president, the Jacksonville Agreement had never been put into effect by some of the coal operators, and most of the operators who signed the agreement went back to work, operating nonunion. By 1930, mine worker membership had dropped to a point where it had no influence, and the organization was struggling to stay alive. The union membership dwindled so much that at one point, John L. Lewis had to borrow money against his home mortgage to keep the UMW going. So, Mr.

3. Damon D. Watkins, *Keeping the Home Fires Burning; A Book About The Coal Miner*, p. 91.

Lewis dispatched Vice-President Phillip Murray to southern Ohio to urge the mine workers on strike in that state to return to work nonunion, and also, to notify them that the mine workers organization had run out of money and could not keep them on strike any longer.

John Owens, former secretary-treasurer of the International Union, was on the payroll of the district as a representative when he got the message from Phillip Murray that the International Union had cut off all support for the district and its officers because of lack of funds. He supposedly remarked to Phillip Murray angrily that he would have nothing to do with the union in the future; that there was no need for them to ask for his support. Mr. Owens later purchased a gasoline station in his hometown of Cambridge, Ohio, and operated the station until 1933, when he was put back on the payroll by Percy Tetlow, provisional president of District Six. During his absence, Mr. Owens was not a member of the mine workers union because he refused to pay dues after he was cut off in 1928. Mr. Tetlow had arranged to get him back into the union and later, Mr. Owens was named the president of District Six and he had his son, Ronald Owens, named secretary-treasurer and his son, Willard, a lawyer, put on the international payroll. Thereafter, this family received total salaries for all three of approximately $100,000 per year for many years from the UMW treasury.

CHAPTER IV

The Roosevelt Era

In November of 1932, Franklin Delano Roosevelt, a Democrat governor, was elected to serve the people of this country as president and was sworn into office on March 4, 1933.

In the 1932 presidential election campaign, Mr. Lewis, a lifelong Republican, lived up to a commitment he had made to President Herbert Hoover and supported him over the successful presidential candidate, Franklin D. Roosevelt. During this campaign, Phillip Murray, vice-president, and Thomas Kennedy, secretary-treasurer of the UMW, supported President Roosevelt. Thomas Kennedy later was elected Lieutenant Governor of Pennsylvania on the Democratic ticket and served under the administration of Governor Earle.

After his inauguration, President Roosevelt summoned labor leaders, including Phillip Murray and Thomas Kennedy, together with some of the most influential business leaders in the nation to the White House for a conference to discuss legislation that would put Americans back to work.

As a result of this conference, President Roosevelt's administration submitted legislation to Congress creating the National Industrial Recovery Act (NIRA). The NIRA, a basic part of Franklin D. Roosevelt's New Deal Program, empowered the president to set up codes of fair competition for various segments of business and industry. It was premised on the federal government's constitutional authority to regulate interstate commerce. This act was passed by Congress and became law on June 16, 1933. Its first and only administrator, appointed by President Roosevelt, was Gen. Hugh Johnson. Included in the NIRA was a significant provision which gave workers the right to bargain collectively with their employers through representatives of their own choosing. These provisions were mandated in the "Blue Eagle" codes that were made applicable in the various industries. This act also provided that employers would not work their employees in their mills, mines, factories, or other establishments in this country for more than eight hours per day and forty hours per week.

As a result of the NIRA being passed by Congress, even though in 1935 it was found by the U.S. Supreme Court to be unconstitutional, by late 1933 and early 1934, unionization proceeded apace in the basic industries, and millions of American workers were taken off the unemployment rolls.

Taking his cue from Section 7A of the NIRA, Mr. Lewis assembled some of his closest associates in the summer of 1933 and sent UMWA organizers throughout the coalfields of the United States, including the southern coalfields; by early fall of 1933, the complete coal industry was organized under the banner of the UMWA. These organizers were not paid employees of the union. The union was practically broke, and the efforts put forth by these organizers were done without pay. In West Virginia, they were under the strict supervision of Van A. Bittner, an international representative. The union's main organizing campaign was conducted in the southern coalfields because most of this area had never been unionized and most companies, until 1932, had required their employees to sign an employment contract which was commonly known throughout the coalfields as a "yellow dog contract." The "yellow dog contract" made employment contingent upon not becoming a member of the union. It had been declared valid by the U.S. Supreme Court in 1917 in a case involving the Hitchman Coal and Coke Company, but was finally outlawed in 1932 by the Norris-LaGuardia Act.

By the fall of 1933, the union had organized the Central Competitive Field, which included Indiana, Illinois, Pennsylvania, Ohio, and Kansas, and all of the nonunion coalfields throughout the United States. This gave the union a combined membership of approximately 475,000 members.

In the fall of 1933, the operators in the bituminous coal industry agreed to meet at the Shoreham Hotel, in Washington, D.C., with the leaders of the UMW to negotiate a collective bargaining agreement covering the Appalachian area and the Central Competitive Field. By September 21, 1933, the coal industry and the leaders of the UWMA had successfully negotiated a wage agreement known as the Appalachian Wage Agreement, covering the whole bituminous coal industry, which became effective October 2, 1933, and expired April 1, 1934.

On March 29, 1934, the union and the operators signed a new agreement, effective April 1, 1934, whose principal provisions were a seven-hour work day, an increase in day wages of

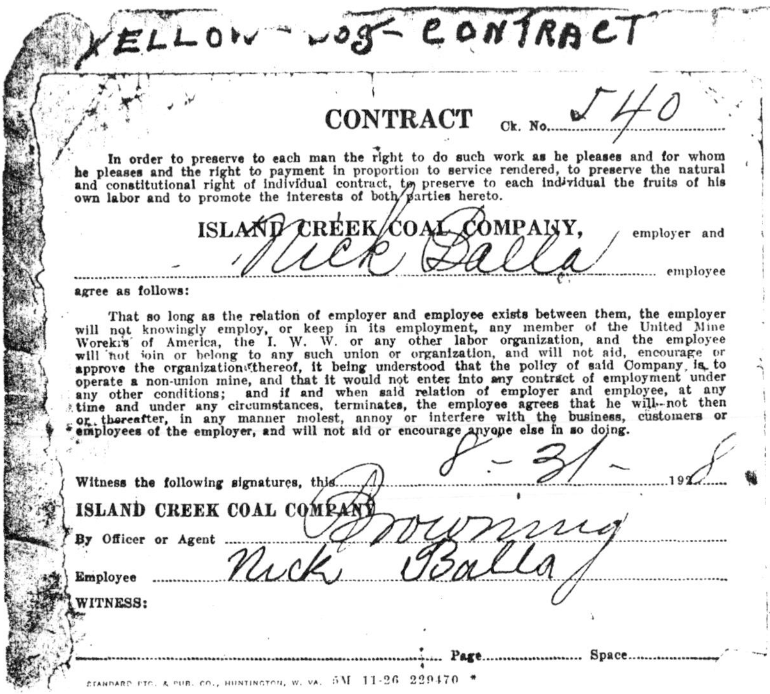
YELLOW-DOG-CONTRACT

CONTRACT Ck. No. 540

In order to preserve to each man the right to do such work as he pleases and for whom he pleases and the right to payment in proportion to service rendered, to preserve the natural and constitutional right of individual contract, to preserve to each individual the fruits of his own labor and to promote the interests of both parties hereto.

ISLAND CREEK COAL COMPANY, employer and

Nick Balla employee

agree as follows:

That so long as the relation of employer and employee exists between them, the employer will not knowingly employ, or keep in its employment, any member of the United Mine Worekrs of America, the I. W. W. or any other labor organization, and the employee will not join or belong to any such union or organization, and will not aid, encourage or approve the organization thereof, it being understood that the policy of said Company is to operate a non-union mine, and that it would not enter into any contract of employment under any other conditions; and if and when said relation of employer and employee, at any time and under any circumstances, terminates, the employee agrees that he will not then or thereafter, in any manner molest, annoy or interfere with the business, customers or employees of the employer, and will not aid or encourage anyone else in so doing.

Witness the following signatures, this 8-31-1928

ISLAND CREEK COAL COMPANY

By Officer or Agent Browning

Employee Nick Balla

WITNESS:

Page Space

STANDARD PTG. & PUB. CO., HUNTINGTON, W. VA. 5M 11-26 229470

forty cents per day, a tonnage rate increase of eight cents per ton and a check-off provision for union dues and assessments.

After the negotiations were concluded, the UMW International Union on July 1, 1934, moved its international headquarters from Indianapolis, Indiana, to the Tower Building in Washington, D.C. Due to the new check-off provision in the new agreement, the union was able to accumulate some money due to its large membership, and as a result of this income, an option was secured to purchase the University Club Building located on the corner of Fifteenth and Eye streets. In the ensuing negotiations with the sellers, a purchase price was secured of $275,000. "The matter was submitted to the International Executive Board for consideration, and the Board by unanimous vote authorized the purchase of the building and its remodeling so that it might be adapted to the office needs and other requirements of the International Union."[1] Looking at

1. International Executive Board, United Mine Workers of America, *John L. Lewis and the International Union, United Mine Workers of America: The Story From 1917 to 1952,* p. 103.

the transaction from an investment standpoint, it was especially commendable because the entire cost of purchase and remodeling was $550,000, while reputable and responsible real estate experts in their conservative opinion advised them that the building, as it stood in 1937, was worth a million dollars.

The years of 1933 and 1934 were the most successful years of John L. Lewis's career as president of the UMW due to his success in organizing the coalfields in the United States under the NIRA. Also, in 1934, he was successful in the San Francisco, California Convention of the AFL in getting the AFL to adopt a resolution pledging its manpower and resources to go out and organize the basic industries of this country, such as rubber, automobile, big and little steel, meat-packing workers, and other nonunion workers throughout the country. A resolution pledging support was passed by a majority vote. However, the AFL's Executive Council arbitrarily refused to act on the resolution adopted in the San Francisco convention, and therefore, took no action on the resolution. In the meantime, the U.S. Supreme Court had, in 1935, declared the NIRA unconstitutional and its regulatory agency was disbanded.[2]

In July of 1935, however, the Congress of the United States enacted the National Labor Relations Act (NLRA) (Wagner Act) which contained a provision identical to Section 7A of the old NIRA. This act was sponsored by the late Sen. Robert F. Wagner of New York, who was instrumental in getting the act through Congress. It became known as labor's Magna Charta and it brought into existence the National Labor Relations Board (NLRB). It was held to be constitutional by the U.S. Supreme Court in 1937.[2]

At the 1935 convention of the AFL held in Atlantic City, New Jersey, Lewis demanded of the AFL officers why there had been no action taken on the resolution that had been passed the year before. A bitter denunciation opposing the resolution was made to the delegates by William Hutcheson, president of the Carpenter's Union. As he denounced John L. Lewis's activities, Lewis jumped to his feet and hit President Hutcheson and knocked him to the floor. The convention adjourned without any further action being taken on the resolution.

2. *Schechter Corporation vs. United States* (1935), 295 U.S. 495.

3. *NLRB vs. Jones & Laughlin Steel Corporation* (1937), 301 U.S. 1.

CHAPTER V

The Birth of the Committee for Industrial Organization

On October 20, 1935, the day after the AFL convention ended, John L. Lewis met with Phil Murray, Tom Kennedy, John Brophy, Sidney Hillman (amalgamated clothing workers), David Dubinsky (women's garment workers), Charles Howard (printers), Thomas McMahon (textile workers), and Max Zaritsky (cap makers), for breakfast at the President Hotel in Atlantic City. "Mr. Lewis told his eight companions, who also had endorsed industrial unionism during the convention, that only they could answer the pleas of the unorganized, that they must not let their defeat obstruct efforts to unionize mass production workers. But the breakfast ended with the advocates of industrial unionism as far from achievement of their primary objective as they had been when the AFL convention opened the previous week."[1] On November 9, 1935, eleven labor leaders met at UMW Headquarters in Washington, D.C., and created the Committee for Industrial Organization (CIO) as an organized bloc inside the AFL dedicated to unionizing mass production workers into industrial-type unions. The CIO was composed of the following eight international union presidents:

1. John L. Lewis, president,
   United Mine Workers of America;
2. Charles P. Howard, president,
   International Typographical Union;
3. Sidney Hillman, president,
   Amalgamated Clothing Workers of America;
4. David Dubinsky, president,
   International Ladies' Garment Workers Union;

1. Melvin Dubofsky and Warren Van Tine, *John L. Lewis, A Biography*, p. 222.

5. Thomas F. McMahon, president,
   United Textile Workers of America;
6. Harvey C. Fremming, president,
   Oil Field, Gas Well and Refinery Workers of America;
7. Max Zaritsky, president,
   Cap and Millinery Department, United Hatters,
   Cap and Millinery Workers' International Union;
8. Thomas H. Brown, president,
   International Union of Mine, Mill and Smelter Workers.

"Unions in America are divided into two general classifications commonly referred to as industrial or "vertical" and craft unions. The first type organizes its members into one union regardless of the type of work done but insists that the laborer shall be paid according to his skill. The second organizes its members into many unions according to their skill thus making in one industry a division of ranks. The background of the industrial type is traceable to continental Europe while the craft union is traceable to the British type."[2]

At this meeting, the officers of the CIO were named. President John L. Lewis of the UWM was named chairman of the committee; Charles P. Howard, president of the International Typographical Union (ITU) was named secretary; and John Brophy, Lewis's one-time opponent for the presidency of the UMWA, was named director to supervise the activities of the committee.

After the formation of the CIO, these eight international unions withdrew their membership from the AFL and in due course, AFL President William Green and the AFL ordered the committee to cease and desist its activities in its organizing drives. However, the committee continued its campaign of organization in the mass production industries.

"Immediately after the committee had been formed, ex-coal miner William Green wrote a letter to Mr. Lewis in which he charged the CIO was a dual organization and that "bitterness and strife would inevitably follow." Mr. Lewis replied:

> My Dear Green:
>
> Mr. Charles P. Howard, secretary of the Committee for Industrial Organization, and other members of the committee have replied categorically and conclusively to the statements contained in your public letter. I associate myself with their replies.

2. Damon D. Watkins, *Keeping the Home Fires Burning; A Book About the Coal Miner*, p. 121.

Now of other things: Your official burdens are great; I would not increase them. I do not covet your office; in proof, I submit the record of years of support of your personal and official fortunes. It is bruited about, however, that your private sympathies and individual inclinations lie with the group espousing the industrial type of organization, while your official actions and public utterances will be in support of their adversaries. Such a policy is vulnerable to criticism and will hardly suffice to protect you against attacks from those who may feel rightfully that more is due them than perfunctory support.

Why not return to your father's house? You will be welcome. If you care to dissociate yourself from your present position, the Committee for Industrial Organization will be happy to make you its chairman in my stead. The honorarium will be equal to that you now receive. The position would be as permanent as the one you occupy. You would have the satisfaction of supporting a cause which you believe inherently, and of contributing your fine abilities to the achievement of an enlarged opportunity for the nation's workers.

Sincerely yours,
(Signed John L. Lewis)

"The former secretary-treasurer of the UMWA refused the CIO post and determined to appeal the AFL's case directly to the UMWA convention to be held in Washington in January, 1936. The following account is from the *Mine Workers Journal:*

President Green addressed the convention for nearly two hours in an endeavor to induce the delegates to reverse the policy they had previously adopted. He should have known better. He should have known, and no doubt he did know, that he would fail. He used everything he had—voice, oratory, argument and gesture. He pleaded, cajoled, intimidated and practically threatened the convention in his effort to carry his point. He even gave a history of his life. But it all got him nowhere. He was "booed" two or three times, but President Lewis stopped the outburst. When it was all over, Green stood a badly defeated and beaten man, and his organization, the AFL, had gone down a little further in the estimation of the 1,700 delegates and the hundreds of spectators who filled Constitution Hall. Surely, it was a humiliating experience for the president of the Federation.

But the climax of the drama came at the conclusion of Green's speech. No sooner had Green said "Thank you" to the convention than President John L. Lewis arose on the stage amid a great demonstration by the delegates standing in their places. They cheered and applauded President Lewis until the windows rattled. When quiet had been restored, President Lewis spoke, slowly, deliberately, but with intense earnestness. He said:

"The President of the United Mine Workers of America will permit the delegates to the Thirty-fourth Constitutional Convention of this organization to render their answer to President Green of the AFL. Let me call upon all delegates in this convention who have changed their minds on this issue on account of the address of President Green to rise to their feet.

"The chair sees two delegates.

"Again, the question recurs upon the fiat of the Executive Council of the AFL, read to this convention as an ultimatum by President Green. It demands that the president of the United Mine Workers of America, with his associates on the Committee for Industrial Organization, like quarry slaves at night, scourged to their dungeon, dissolve, disband, cease and desist with reference to the CIO. Let those delegates of this Thirty-fourth Constitutional Convention who believe that the president of the United Mine Workers of America should comply with that request rise to their feet.

"The chair sees one delegate arise.

"Again, let those delegates of this convention who believe that the policies enunciated by this convention should be carried out by the president of their organization and his associate officers rise to their feet."

(The delegates arose and applauded.)

"President Green, you have received the answer of the United Mine Workers of America to your ultimatum. It is not for the president of the United Mine Workers of America to amplify with mere words an expression of a principle and a conviction so deep seated, so pronounced, and so traditional as exists with reference to this question.

"You come as an ambassador from another organization to the United Mine Workers of America. I hope, sir, that you have been treated with all the courtesies and honors due an ambassador, but you have and you may carry back to your organization, the answer of the United Mine Workers of America that has just been given by this convention!"

"Never before in any convention of the UMWA did anyone witness a scene like this one. Every man in the house was on his feet, clapping hands, cheering and through the medium of noise giving evidence of the depth of his enthusiastic approval of President Lewis' words. In the deep seriousness of the moment, President Lewis, his jaws set and his face a picture of determination and finality, faced the frenzied enthusiasm of the vast crowd which was thus pledging the support of more than half a million coal miners to their leader. Green, red of face and almost voiceless after his long speech, sat there with his gaze fixed upon the cheering throng. What was passing through his mind can only be surmised. After several minutes of the demonstration, the crowd became quiet, and Green left the Hall, and the routine work of the convention was resumed. But no one who was fortunate enough to be present at that historic occasion will ever forget the scene and its significance."[3]

3. International Executive Board, United Mine Workers of America, *John L. Lewis and the International Union, United Mine Workers of America: The Story From 1917 to 1952*, pp. 71-72; 74.

CHAPTER VI

The Birth of the Congress of Industrial Organizations

The years of 1936 and 1937 were two of the most outstanding years in the career of John L. Lewis. In the early part of 1936, the money interests of Wall Street organized what was at that time called the famous "Liberty Leaguers." This group pledged itself and its finances to defeat President Roosevelt in the 1936 presidential campaign for his second term. President Roosevelt's campaign was having difficulty, as a result of the Liberty League, in raising campaign funds, and the UMWA was the largest contributor to the Roosevelt campaign.

The UMW contributed $500,000 and pledged its support to the reelection of President Roosevelt. Also, the UMW formed Labor's Non-Partisan League (the political arm of the CIO) to help President Roosevelt's campaign for his second term.

In the November General Election, President Roosevelt defeated his Republican opponent, Alf Landon of Kansas, by carrying every state in the Union with the exception of Maine and Vermont.

When Mr. Lewis was elected president of the UMW in 1920, and the constitution was changed to prohibit members of the Communist party, International Workers of the World, and other subversive organizations opposed to Mr. Lewis in his ascendancy to the presidency of the UMW, were barred from membership in the UMW union. The men expelled from the UMW union were: John Brophy of central Pennsylvania, president of District Two, who ran against Mr. Lewis for president of the UMWA; Adolph Germer, a coal miner from Illinois; Powers Hapgood, the son of a millionaire shoemaker and a graduate of Yale who was working in the mines of Indiana; Allan Haywood of Pennsylvania, who was a leader in the National Miners Union (a dual organization to the UMW), and

Alexander Howat of Pittsburgh, Kansas, former president of District Fourteen.

Mr. Lewis recognized their leadership ability and organizational skills; reinstated their membership in the UMW; and placed them in positions of leadership in the CIO throughout the country. The rubber industry was the first industry organized by the CIO, and Allan Haywood was placed in charge of that campaign. In less than four months, the rubber industry was completely organized, and Sherman J. Dalrymple, a rubber worker, was named as the first president of the United Rubber Workers Union (URWU). Adolph Germer was placed in charge of the organization in the state of Michigan, and he began a campaign to organize General Motors. In a matter of a few weeks after this campaign began, the employees of General Motors Corporation (GMC) started a sit-down strike in the plants of General Motors and refused to leave the plants. In the meantime, the employees urged Mr. Lewis to come to Detroit to help negotiate a contract with the GMC. John L. Lewis held several conferences with the officers of GMC and Gov. Frank Murphy of Michigan. As a result of these conferences, General Motors signed a contract, the first labor agreement signed in the auto industry; with Mr. Lewis on February 11, 1937, covering the employees of GMC.

"The General Motors stay-in strike lasted forty-four bitter, wintry days. The company's negotiators were adamant in their refusal to sign a contract, and their attitude was reflected in the actions of their underlings who made life as miserable as possible for the freezing men and women in the plants and on the picket lines.

"President Roosevelt and Secretary of Labor Perkins tried unsuccessfully to conciliate the dispute. William Green tried to break up the negotiations. One of the GM negotiators, William Knudsen, produced a telegram from Mr. Green at one of the meetings, which asked that the AFL be made a party to the talks. Mr. Green's request was overruled when Mr. Lewis suggested that Haile Selassie be invited to participate because he represented the same number of GM workers as the AFL.

"But it was the patience of Governor Frank Murphy of Michigan, who refused to listen to "public clamor" for the use of troops against the strikers, which kept the GM officials in contact with Mr. Lewis until they finally realized that the workers were not to be defeated.

"The contract granting union recognition was signed in Detroit, February 11, 1937. In a later description of the end of the strike, Mr. Lewis said:

At three o'clock in the morning on one of the high floors of the Statler Hotel, Mr. Knudsen, the president; Donaldson Brown, chairman of the finance committee of the board, and John Thomas Smith (the GM negotiating committee) walked into my room, when I was in bed. And they had on their overcoats, and they had their hard hats in their hands, and their gloves on because the room was cold. I didn't get up, and they said that they would sign the contract at eleven o'clock that morning in Governor Murphy's office. And they did.

"The financial world was stunned on March 2, 1937, when it read in the newspapers that the United States Steel Corporation had, without a struggle, signed a wage agreement with the CIO. The signs all had pointed toward a long, bitter strike.

"The contract was the culmination of a long series of conferences between Mr. Lewis for the CIO and Myron C. Taylor for U. S. Steel. In a press statement issued on the day the contract was signed, Mr. Lewis called it 'a fine example of an intelligent approach to a great economic problem.'

"Shortly after U. S. Steel had been signed up by the CIO, a month-old strike of 65 thousand workers at several plants of the Chrysler Corporation ended. Walter P. Chyrsler, Mr. Lewis and representatives of the UAW signed a contract after several days of conferences. Much credit for the successful meetings went to Labor's good friend, Governor Murphy. The agreement ended the company-union, industrial spy system which Chrysler had used in its vain attempts to defeat the auto workers."[1]

Republic Steel was headed by Tom Girdler as president, who vowed he would never sign a union contract. On Memorial Day, May 1937, the families and workers of Republic Steel gathered by the thousands in south Chicago for a Memorial Day celebration. During the celebration, there were hundreds of Chicago policemen present to keep the peace, and eventually fighting broke out. After calm had been restored, ten strikers had been killed and nearly one hundred wounded.

As a result of this massacre, John L. Lewis, as president of the CIO, called upon President Roosevelt to investigate the activities of the police department at the Memorial Day Massacre. President Roosevelt responded by quoting Shakespeare, "A plague on both your houses,"—blaming both the steelworkers

1. International Executive Board, United Mine Workers of America, *John L. Lewis and the International Union, United Mine Workers of America: The Story From 1917 to 1952*, pp. 90-91.

and Republic Steel for the massacre. Mr. Lewis responded by paraphasing Shakespeare, "It ill behooves one who has supped at labor's table,"—meaning that President Roosevelt had enjoyed the support of these workers during his reelection campaign. Mr. Lewis called upon Sen. Bob LaFollette, chairman of the Civil Liberties Committee of the U.S. Senate, to investigate the activities of the police department of the city of Chicago in the Memorial Day Massacre. The committee went to Chicago and spent considerable time on the investigation, and as a result, blamed the Chicago Police Department for the massacre. John L. Lewis, as president of the CIO, urged President Roosevelt to have Mayor Kelly prosecute the police officers who were responsible for this massacre. The mayor of Chicago ignored this plea. Since President Roosevelt would not pursue this matter with the mayor of Chicago, this was the beginning of the split between Mr. Lewis and President Roosevelt.

In January of 1938, the UMWA International Convention was held in the Rialto Theater in Washington, D.C. This was the result of the Daughters of the American Revolution (DAR) organization refusing to let the UMW hold their convention in Constitution Hall, where the convention had been held in 1936. The reason for the refusal by the DAR organization was because, in 1936, the union did not segregate the blacks and whites, and it was the first time in the city of Washington that the whites and blacks sat in meetings together without any segregation.

After the January 1938 convention, Mr. Lewis began preparing for the convention of the CIO to be held in Pittsburgh, Pennsylvania, at the Grotto Hall, 107 East Montgomery Avenue, November 14, 1938. I was an official delegate to this convention and occupied a table in the convention with Mr. Heywood Broun, who was a nationally known syndicated writer at that time and International President of the Newspaper Guild. Mr. Broun was a faithful, loyal supporter of John L. Lewis, and I was very much impressed with his attitude toward labor unions which we discussed during the whole week I attended the convention.

In this convention, it was decided that the CIO would be made a permanent organization and the name was changed from the Committee for Industrial Organization (CIO) to the Congress of Industrial Organizations (CIO), and thus they were proclaimed the national organization by act of the convention.

I felt highly honored to have been a part of forming the Congress of Industrial Organizations.

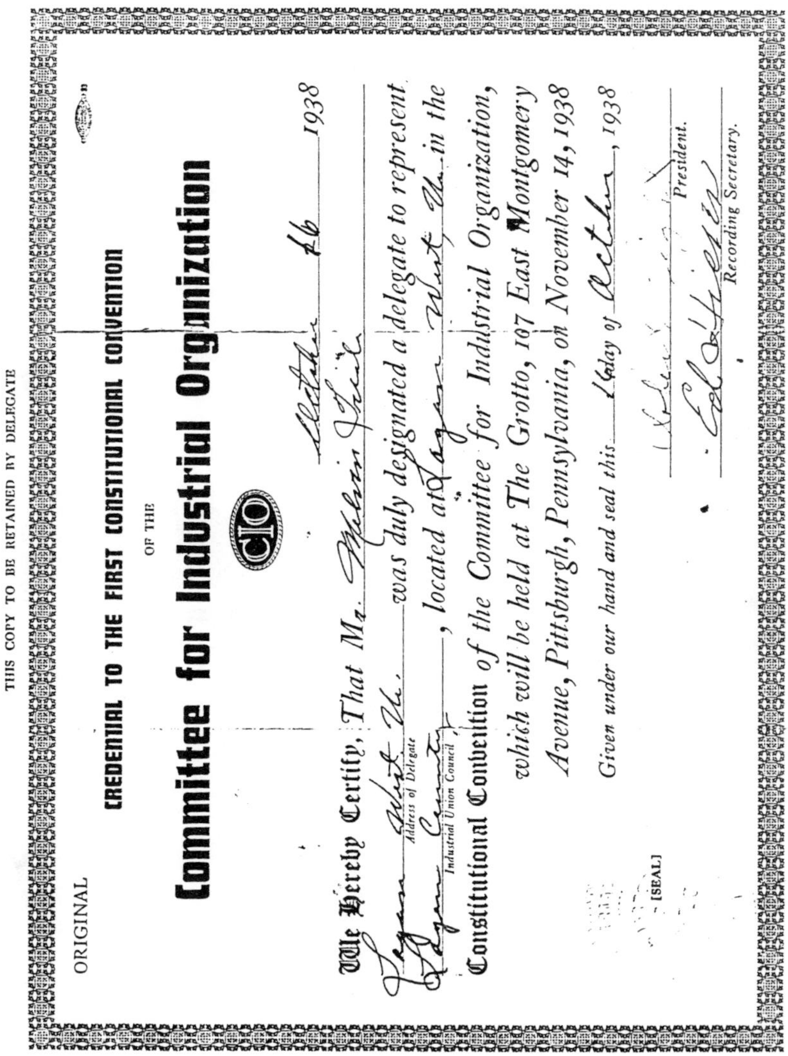

THIS COPY TO BE RETAINED BY DELEGATE

ORIGINAL

CREDENTIAL TO THE FIRST CONSTITUTIONAL CONVENTION

OF THE

Committee for Industrial Organization

CIO

October 26 1938

We Hereby Certify, That Mr. Melvin Friel

Logan West Va.
Address of Delegate

was duly designated a delegate to represent

Logan County
Industrial Union Council

, located at Logan West Va. in the Constitutional Convention of the Committee for Industrial Organization, which will be held at The Grotto, 107 East Montgomery Avenue, Pittsburgh, Pennsylvania, on November 14, 1938

Given under our hand and seal this 26 day of October, 1938

[SEAL]

President.

Recording Secretary.

The officers of the newly created national organization were John L. Lewis, president; Sidney Hillman, vice-president; Philip J. Murray, vice-president; and James B. Carey, president of the United Electrical Workers Union, as secretary. John Brophy was made director of organization.

CHAPTER VII

UMW Golden Anniversary (1890-1940)

In the spring of 1939, negotiations between the UMW and the coal operators were being held to negotiate a new wage agreement for the industry. With the economy being down, the union did not press the coal industry for any wage increases. However, as a result of these negotiations, the operators agreed to the union's demand for a union shop clause in the new agreement. This was the only concession that the union was able to extract from the coal industry. The union shop clause written into this labor agreement was the first such provision incorporated in a labor agreement in this country. This provision required the employer to employ members of the UMW organization as a condition of employment in the industry, and if an employee did not agree to join the UMW, he was terminated from his employment.

At the conclusion of the negotiations of the labor agreement, it was purported by some politicians and labor leaders that the rift between President Roosevelt and John L. Lewis had become more apparent, and it was alleged in the political circles that the reason for this rift was due to John L. Lewis's demand that he be President Roosevelt's running mate in the 1940 campaign, and that President Roosevelt had rejected Mr. Lewis's request. There is no truth to this matter. Mr. Lewis, under no circumstances, was interested in getting into politics. However, he did prevail upon the supporters of President Roosevelt of the Democratic party to consider as President Roosevelt's running mate Lieutenant Governor Thomas Kennedy of Pennsylvania, who was also supported by Governor George Earle of Pennsylvania, and the Democratic organization of Pennsylvania. This recommendation was turned aside, and President Roosevelt's vice-presidential selection was Henry A.

Wallace, who was in President Roosevelt's Cabinet as the Secretary of Agriculture.

During the early part of 1939, by proclamation, President Roosevelt set up the National Defense Program, getting ready for World War II. The director of the National Defense Program, who was named by the president, was Sidney Hillman, the vice-president of the CIO and president of the Amalgamated Clothing Workers Union. At the CIO convention in November of 1939, in San Francisco, California, Mr. Lewis offered the convention his resignation as president. This was due mainly because of his concern that some of the leaders in the CIO were out-and-out Communists and followed the Communist line. Such leaders were Harry Bridges of the Longshoremen's Union; Michael Quill of the Transport Workers Union; Sam Woolchock of the United Wholesale and Restaurant Workers Union; Joe Curran of the Maritime Union; Lee Pressman, General Counsel of the CIO; and Len De Caux, editor of the *CIO News,* the news publication of the CIO. However, after some persuasion by Vice-President Murray and Vice-President Hillman, Mr. Lewis agreed to remain as president of the CIO and was reelected to another term at the convention.

On the fourth Tuesday of January in 1940, which fell on January 30, 1940, the UMW held their Golden (Fiftieth) Anniversary convention in the city of Columbus, Ohio, where the UMWA was founded on January 25, 1890. Mr. Lewis invited as the guest speakers for the convention Sen. Burton K. Wheeler of Montana, the United States Senate's most outstanding isolationist and Gen. Hugh Johnson, the former administrator of the NIRA, and also one of America's ardent isolationists. Both Senator Wheeler and General Johnson opposed President Roosevelt for a third term. They were the only speakers invited to the convention. At the opening session of the national convention, Mr. Lewis, in his opening remarks, stated that if President Roosevelt was a candidate for a third term, then, in his opinion, President Roosevelt would go down in ignominious defeat. I had been attending the conventions since 1936, and this is the first time I recall that President Lewis did not receive any applause. In fact, you could have heard a pin drop after he made this announcement regarding President Roosevelt. After his speech, the delegates from all over the convention floor went to the microphone and stated their objections to Mr. Lewis's remarks against President Roosevelt.

Later, when the convention had settled down, the next order of business was a national hook-up on the National Broadcasting

Corporation radio station carrying the speeches of Thomas Kennedy, secretary-treasurer; Phillip Murray, vice-president; and President John L. Lewis, commemorating the golden anniversary of the UMWA. During Mr. Lewis's speech, a large red flag dropped from the ceiling and some of the delegates did not recognize that this was the Communist flag. Mr. Lewis was handed a note during his speech regarding what had happened, and before he concluded, he demanded the authorities of the city of Columbus, Ohio, to make a formal investigation and prosecute those responsible. Later, it was reported that the Screen and Stage Workers Union affiliated with the AFL was responsible; but there was no evidence submitted that indicated that they had had anything to do with it.

After the convention in Columbus, and later in the summer, the Democratic National Convention nominated President Roosevelt for a third term and Henry A. Wallace as his running mate for the vice-presidency. The Republican National Convention nominated Wendell L. Wilkie, a utility executive from Indiana, for president.

On Monday, October 21, 1940, it was announced by Mr. Lewis's office that Mr. Lewis would make a major speech over the three major broadcasting networks, NBC, CBS and Mutual Broadcasting Company, regarding the forthcoming presidential election in November between President Roosevelt and Wendell L. Wilkie. "Friday evening, October 25, found 25,000,000 to 30,000,000 Americans listening to the sonorous tones of an undreamed-of Lewis castigation of the Chief Executive. Over America's three major radio broadcasting chains involving 322 radio stations, Lewis hurled every weapon in his arsenal at Roosevelt. Throughout Lewis's attack upon Roosevelt he constantly hammered upon one theme. Roosevelt means War! War! War!"[1]

Lewis stated:

"Those who hear these words, and who have studied the public addresses of the President, from his Chicago 'quarantine speech' to his Charlottesville 'stab in the back' address, and thence to Dayton and Philadelphia, will understand his motivation and his objective. It is war. His every act leads one to this inescapable conclusion. The President has said that he hates war and will work for peace, but his acts do not match his words. I am opposed to any involvement of our country in foreign wars. I believe that every thoughtful and normal citizen

1. Saul D. Alinsky, *John L. Lewis: An Unauthorized Biography*, p. 188.

is similarly opposed. They are willing, and I am willing, to contribute everything for any necessary defense of our geographical integrity, our families, our possessions, our liberties, and our lives.

"The present concentration of power in the office of the President of the United States has never before been equaled in the history of our country. His powers and influence in this republic are so far-reaching that they intimately and vitally affect the lives and fortunes of every citizen. In like measure, they may affect the lives and fortunes of other nations and their populations.

"How startling, therefore, is the spectacle of a President who is disinclined to surrender that power, in keeping with the traditions of the republic. The suggestion of a third term under these conditions is less than wholesome or healthy. Personal craving for power, the overweening abnormal and selfish craving for increased power, is a thing to alarm and dismay.

"Power for what? Personal and official power to what end? In all history, the unwarranted exercise of continously vested authority has brought its train of political and social convulsions for which humanity has paid an appalling price in loss of liberty, in disorder, tragedy, and death.

"America needs no superman. It denies the philosophy that runs to the deification of the state. America wants no royal family. Our forebears paid the price in blood, agony, privation, and sorrow, requisite for the building of this republic. Are we now to cast away that priceless liberty, which is our heritage? Are we to yield to the appetite for power and the vaunting ambitions of a man who plays with the lives of human beings for a pastime?

"I say 'No,' and whether I stand alone, or whether I am sustained, as I think I will be, by the overwhelming number of American citizens, I should retain these convictions. It is time for the manhood and the womanhood of America to assert themselves. Tomorrow may be too late.

"President Roosevelt is asking the American people to contribute to him at least four more years out of their individual lives. What will be done with those lives and this nation in the next four years, and how does he propose to do it? He has not said, and he asks from the people a grant of discretionary power that would bind him to no course of action, except the unpredictable policies and adventures which he may later devise.

"After all, Americans are not a nation of guinea pigs, constantly subject to the vicissitudes of the economic and political experiments of an amateur, ill-equipped practitioner in the realm of political science.

"If not Roosevelt, whom do I recommend to do the job of making secure our nation and its people? Why, of course, I recommend the election of Wendell L. Wilkie as the next president of the United States.

"He is a gallant American. He has opened his heart to the American people. He is not an aristocrat. He has the common touch. He was born to the briar and not to the purple. He has worked with his hands, and has known the pangs of hunger. He has had experience in various fields of American enterprise, and is an administrator and an executive.

"He is strong enough to enlist the services of other strong men to do the job of saving our nation, whether from attack by external foes, or disintegration from disunity within.

"It is obvious that President Roosevelt will not be re-elected for a third term, unless he has the overwhelming support of the men and women of labor. If he is, therefore, re-elected, it will mean that the members of the Congress of Industrial Organizations have rejected my advice and recommendation. I will accept the result as being the equivalent of a vote of no confidence and will retire as President of the Congress of Industrial Organizations, at its convention in November. This action will save our great movement, composed of millions of men and women, from the embarrassment and handicap of my leadership during the ensuing reign of President Roosevelt.

"To the leaders of the CIO, its executives, staff officers, and field representatives—I know and have worked with each of you. Upon some of you, I have bestowed the honors which you now wear. Through the years of struggle, you have been content that I should be in the forefront of your battles. I am still the same man. Sustain me now, or repudiate me. I will not chide you and will even hope that you will not regret your action.

"To the mine workers of the nation who know me best, and who have always been the shock troops in the forward march of labor, I say it is best for you, and for those you love to help oppose the creation of a political dictatorship in free America.

"To the steel workers, the automobile workers, the shipbuilders, the maritime workers, the lumber workers in the far northwest, the textile workers, the white-collar workers, and the men and women of labor in the miscellaneous industries, I say I have worked for you and have fought for you. Believe me now, when I say that your interest, and the interests of the families you support, lie in the acceptance of the truth of the words I speak tonight."

"Union men and women wept with bitter disappointment at this break between their two idols. Liberals and many adherents of the CIO cursed Lewis for what they felt was a major betrayal. Ten days later Franklin Delano Roosevelt stepped into the spotlight of history as America's first third-term president, and John Llewellyn Lewis stepped in the long twilight."[2]

A large majority of the coal miners of the United States overwhelmingly supported President Roosevelt's reelection. The mine workers slogan was that they were for President Roosevelt for president of the United States and for President Lewis as president of the UMW. In the meantime, many of the leaders of the CIO were very critical of Mr. Lewis for supporting Wendell L. Wilkie and so expressed themselves in the public press, and some were critical of Mr. Lewis because his speech opposing President Roosevelt for a third term was carried in its entirety in the *CIO News*, a CIO publication.

2. Saul D. Alinsky, *John L. Lewis: An Unauthorized Biography*, pp. 188-191.

CHAPTER VIII

Lewis Resigns as CIO President

After the election, Mr. Lewis made preparations for the CIO convention in Atlantic City beginning on Monday, November 18, 1940, in the Chelsea Hotel. Prior to Mr. Lewis's departure for Atlantic City, all of his district presidents, international board members, and all district and subdistrict representatives were ordered by Mr. Lewis to attend the convention.

Myself and the officers and representatives of District 17 left by train for Atlantic City on Sunday, November 17, 1940. The officers and members of the CIO were mostly housed in the Chelsea Hotel in Atlantic City on the Boardwalk. On Monday morning, November 18, at 10:00 a.m., Mr. Lewis opened the convention and proceeded to make his opening statement to the delegates. As he began to talk, a spontaneous demonstration developed with delegates wearing all kinds of convention badges with slogans on them such as "We Want Lewis," "Forward with the CIO," and delegates, using horns, other tin noisemakers and banners paraded around the convention hall. Mr. Lewis stood silently and watched the demonstration. After a while, the demonstration ended, the delegates took their seats, and Mr. Lewis proceeded with his opening remarks. He reiterated to the delegates that he would not be a candidate for the office of president. Once again, a demonstration started and lasted for approximately forty minutes. The delegates shouted "We Want Lewis! We Want Lewis!" Finally, Mr. Lewis brought the convention back to order and proceeded with his remarks. Mr. Lewis pointed out that the members of the CIO did not have any confidence in his leadership and repudiated him because of his opposition to the reelection of President Roosevelt. He also pointed out to the delegates that many of the leaders of the CIO had attacked him personally in the public press because of his support of Wendell L. Wilkie. He pointed out that he had received telegrams of protest from the members of the CIO and that many of the officers of the CIO unions had been very critical of him because his speech opposing President Roosevelt for a third term had been printed in the *CIO News*. Mr. Lewis pointed out that he was president of the CIO, and all major addresses made by the president of the CIO in the past had always gotten full coverage in the *CIO News*, and he did not see anything wrong with having his speech against President Roosevelt carried in the CIO press. A demonstration broke out again and lasted for ten or fifteen minutes. The delegates shouted and sang, "Onward with Mr. Lewis and the CIO." These demonstrations were highly organized and well handled. Mr. Lewis's biggest critics in the CIO such as Harry Bridges, president of the Longshoremen's Union; Michael Quill, head of the Transport Workers Union; Joseph Curran of the Maritime Union; Lee Pressman, general counsel of the CIO; and many others who followed the Communist line, were the principal organizers of these demonstrations. After the last demonstration had settled down, Mr. Lewis bellowed into the microphone, "Your cheers and demonstrations do not enthuse me, nor your curses and criticism of my leadership deter me," and vowed that he would continue to be active in his efforts to organize the unorganized workers of this country. He admonished the delegates and pointed out that under his leadership, the Congress for Industrial Organizations in four short years had come a long way; that the CIO had organized all of the mass production industries, in addition to establishing CIO state associations and central points of the CIO in all of the larger cities of this country, with a combined membership of more than four and one-half million members. He continued, and said of the efforts of the CIO organizing drives in the mass production industries had forced the leadership of the AFL to start an organizing campaign to improve its position in the labor movement. In other words, the AFL had to go back to work and organize unorganized workers. He concluded by repeating his opening statement that he would not be a candidate for reelection.

After the convention sessions were over on Monday and Tuesday evening, Mr. Lewis, Thomas Kennedy, secretary-treasurer of the UMWA; and Sidney Hillman of the Amalgamated Clothing Workers Union; had several meetings with Phillip J. Murray, who was then president of the United Steel Workers Union. They pressed him by walking with him on the Boardwalk in the evening and making every effort to convince him that he should accept the presidency of the CIO, succeeding

Mr. Lewis. After they found they could not convince Phil Murray to change his mind, they immediately called Father Charles Owen Rice of Pittsburgh, an outstanding Catholic proponent of labor unions and very close friend of Phil Murray, to come to Atlantic City. Upon his arrival, Father Rice was asked by Mr. Lewis, Mr. Kennedy and Mr. Hillman to see if he could convince Phil Murray to accept the leadership of the CIO. By Wednesday evening, Phil Murray had agreed with Father Rice, and he informed Mr. Lewis, Mr. Kennedy, Mr. Hillman and many others that he would accept the presidency of the CIO.

At the session on Thursday morning, the nomination and election of CIO officers took place. Mr. Lewis rose and took the microphone and nominated Phil Murray, his closest friend and most trusted lieutenant, for the presidency of the CIO. In his nominating speech, Mr. Lewis gestured with his right arm and said, "I am for Phil Murray when he's right and I am for him when he's wrong and I place his name in nomination for the presidency of the CIO," and he pledged his unqualified support to his lifelong friend. At that time, a spontaneous demonstration started again, and it lasted for about sixty minutes.

After Mr. Lewis's nominating address, Sidney Hillman of the Amalgamated Clothing Workers Union, seconded the nomination of Phil Murray and various others also seconded the nomination. Before the close of the session on Thursday morning, Phil Murray was elected president of the CIO; Sidney Hillman was reelected vice-president; James B. Carey of the Electrical Workers Union was reelected secretary-treasurer; and John Brophy was reelected director of organization.

The convention adjourned on Friday afternoon, and the District 17 UMWA delegation boarded a train on Friday evening and arrived in Charleston on Saturday morning about 9:00 a.m. The newspaper boys were selling extra editions of the *Charleston Gazette* with the headline, "Lewis Breaks With Murray." We were astonished to see this headline because on Thursday, Mr. Lewis had pledged his wholehearted support to Mr. Murray. This news of a breakup between the two, announced before they had even left Atlantic City, shocked the labor movement throughout the country.

CHAPTER IX

UMW Strike at Widen—1941

After Mr. Lewis returned to Washington, he began preparations for the 1941 coal wage negotiations. The Appalachian Coal Wage Agreement would expire on March 31, 1941. Negotiations between the coal operators and the UMW began early in 1941 in New York City at the Roosevelt Hotel, where all of the coal wage negotiations were conducted.

"The bituminous negotiations of that year were long and bitter. They opened on March 11 and the final contract—another battle won by the UMWA—was not signed until July 5. The UMWA, with elimination of the 40-cent differential, became the first union to wipe out the wage differential between the North and South. The universal basic daily wage became $7, a token vacation payment of $20 was obtained, and the union shop became universal in the commercial mines, even in Harlan County.

"After three weeks of unsuccessful bargaining in March, the miners stopped work. Dr. John R. Steelman, head of the U. S. Conciliation Service, intervened and found that the Southern operators were balking at the UMWA proposal, supported by the Northern Appalachian operators, that the 40-cent differential be eliminated.

"On April 11, the Southerners withdrew from the conference and later appealed to President Roosevelt to have their case referred to the National Defense Mediation Board.

"On April 19, Secretary of Labor Frances Perkins recommended that mines belonging to operators with whom the UMWA had reached an understanding should reopen, but both operators and the union rejected her request.

"Two days later, the President recommended that talks be resumed. The Southern operators went back to New York City, negotiated briefly and returned to Washington.

"As a result of this second walkout, Secretary Perkins certified the case to the National Defense Mediation Board. A panel of the board recommended that the mines reopen on April 28 on the basis agreed to by the Northern operators, eliminating the differential.

"The Northern operators signed a temporary agreement on that basis on April 29; the Southerners again balked and the mediation board reopened hearings. The board findings were made public by Chairman William H. Davis, and in the main, they were favorable to the UMWA.

"The Northern Appalachian operators signed the new agreement on June 19 and the Southern operators signed on July 5."[1]

During the negotiations with the Appalachian coal operators in early February or March, Mr. Lewis, Vice-President Murray, Secretary-Treasurer Kennedy and President of District 17, Van A. Bittner, held a conference with Ben Fairless, Chairman of the Board of United States Steel Corporation, in New York to explore the advisability of Mr. Fairless signing a contract with the UMWA covering the mine workers in U.S. Steel's coal mines, commonly known as captive coal mines. Mr. Fairless pointed out, that when the union could prove to him that the UMWA represented a majority of U.S. Steel's mine employees, then he would sit down and negotiate with them.

In the meantime, coal wage negotiations continued with the Appalachian operators, and during a break in negotiations, Mr. Van A. Bittner, President of District 17, UMWA, came back to Charleston and held a luncheon meeting of the representatives of District 17 in the Daniel Boone Hotel ballroom. Mr. Bittner discussed with the representatives of District 17, which at that time covered all of West Virginia, with the exception of District 31, which was headquartered in Fairmont and covered northern West Virginia, some of the issues that were being discussed with the Appalachian coal operators. He pointed out that the largest nonunion commercial coal operator in West Virginia was Col. J. G. Bradley, who at that time was president of the West Virginia Coal Association, and who owned ninety percent of the acreage in Clay County, West Virginia, where his nonunion operation was located, known as the Elk River Coal and Lumber Company. In addition to the coal mines, he operated a large lumber company together with his

1. International Executive Board, United Mine Workers of America, *John L. Lewis and the International Union, United Mine Workers of America: The Story From 1917 to 1952,* pp. 134-35.

own railroad which was known as the Gauley River Railroad. Mr. Bittner pointed out that as long as there was a nonunion operator in West Virginia, it would always be a threat to organized labor in West Virginia, and particularly, the UMW. He also pointed out that he wanted the Elk River Coal and Lumber Company closed down at the expiration of the Appalachian Contract on April 1, 1941, the same as other mines in the Appalachian area. He put no limitations on getting this job done. After the representatives returned home from the meeting in Charleston, they began an organizing campaign to carry out the wishes of Mr. Bittner. The representatives of the union who were in charge of the various coal areas in District 17 were as follows: Othar Ballard, Kanawha District; Ermal Hodson, New River and Winding Gulf Districts; Lawrence Hindle, Greenbrier District; Millard Cassidy, Logan District; and Jessie Aquino, Pocahontas and Tug River districts.

They immediately began recruiting coal miners from all of these different areas, and it was agreed that they would meet about two o'clock in the morning on April 1, 1941 on top of Widen Mountain and would, at that time, determine the number of miners that were to be in the picketing caravan. Approximately three thousand were recruited. When they got to the vicinity of Widen, they decided they would go down into the town of Widen itself. The caravan of miners proceeded down into the town about four o'clock in the morning. They circled the whole town and went to the lamphouse at the mine to make sure that nobody would report for work. Mr. Bradley's mine was closed down and did not work for several days. This was the first time in the history of the Elk River Coal and Lumber Company that Colonel Bradley's mine was shut down by UMW pickets.

In the meantime, Mr. Bradley closed his company store, and the union pickets had to go all the way to Clay, about twenty miles over the mountain from Widen, to get something to eat, second, most of the businesses in Clay would not sell them any food or anything to drink. As a result of the union's people not being able to get something to eat or drink, some of them left the scene and drove back home. This left the union with a small group to try to keep Widen on strike during the wage negotiations.

One evening around five o'clock, the sheriff of Clay County, who was the tipple foreman for Elk River Coal and Lumber Company, appeared on the scene with about seventy-five deputies whom he had deputized to serve as special deputies. These "special deputies" had yellow ribbons around their left arms to indicate they were special deputies. Each one of them had in his possession a brand new pick handle that had been given to him by the company. The sheriff had the original copy of a court injunction issued by Judge Fisher of Clay County, and he read the injunction to the union pickets. After he had read the injunction, he served three or four copies on the leaders of the UMWA pickets and ordered them out of the town of Widen immediately. Since the UMWA pickets had dwindled down to practically nothing, they had no choice but to leave town because they all knew that if they didn't proceed and leave Widen, that the fellows the sheriff had deputized would use those pick handles. The deputies had already started calling the pickets names and ordered them out of the town of Widen. So, the pickets obeyed the instructions of the sheriff. When they got to the top of Widen Mountain, they decided to settle there until they could have more UMW picket reinforcements. The next day, they were reinforced with about three or four hundred UMW pickets. While they were on top of Widen Mountain with their new pickets, the deputies the tipple foreman had deputized, parked a company truck crossways in the road going up the mountain and began shooting at us with high-powered rifles. Some of the UMW pickets had high-powered rifles and went down toward where the company deputies were firing. As a result of these shootings, the UMW appealed to Governor M. M. Neely for protection. Governor Neely and Superintendent of Public Safety, Colonel Crumpacker, came to Widen. Both of them talked to the UMW pickets on top of the hill, and then they went back down into Widen to check on the newly made deputies by the sheriff of Clay County. They were gone about two hours, and when they came back to the top of the hill, Governor Neely and Colonel Crumpacker both addressed the group on top of the hill and stated that they would send the state police to see that law and order was carried out.

In the meantime, while Colonel Crumpacker was talking, he became ill, and the governor and his party had to leave because of Crumpacker's illness. Upon arrival in Charleston, Crumpacker was taken to the hospital, and the next day he died of a heart attack. It was widely rumored for many months thereafter, that Colonel Crumpacker had been poisoned while he and the governor had been in Widen. Colonel Crumpacker was the former sheriff of Mercer County before being named the Superintendent of the Department of Public Safety, and he was a member of the Brotherhood of Railroad Trainmen in Princeton. This strike lasted for approximately six weeks.

Later, another incident occurred in Widen where two UMW pickets were shot by the deputies with 30-30 high-powered rifles. One of the fellows that was shot was a Mr. Farley, a young man of twenty-one years of age from Sophia in the Beckley area. The other person shot by the thugs was a UMW picket by the name of Robinette, who was from Stone, Kentucky. He had been hit in the tip of his right elbow and his right hipbone. He had been hunkered down behind a tree when he was shot. The other UMW pickets missed Robinette and Farley and went looking for them. They found Farley lying down the hill near where the company truck was located, and he was taken to Montgomery Hospital at Montgomery, West Virginia. He died en route to the hospital. Robinette was picked up in the woods about a half mile from where the UMW pickets were stationed. He was taken to the Charleston General Hospital in Charleston for treatment.

As a result of the activities on the morning of the first of April and the shooting incidents that followed, the UMW was able to continue its organizing campaign and petitioned the National Labor Relations Board (NLRB) for a representation election at Widen. In the meantime, all of the pickets were withdrawn from Widen, and an election was held by the NLRB sometime in the late summer. The UMWA lost the election. The Widen League of Miners, a company union, was declared the winner.

Afterward, the UMW continued to keep two organizers in Widen for the purpose of organizing the Widen miners. The organizers were Ezra Foley and Ransom Kirk. They spent most of their time in Widen signing up employees of the Elk River Coal and Lumber Company who desired to join the union. The NLRB announced that the UMW had sought another election in the early part of 1942 involving the Widen miners, and the date for the election was set. About a week before the election, Foley and Kirk were down at the YMCA in Widen, and a Mr. Groves, who was the mine foreman at the Widen mine, and his son, who was on furlough from the army, came to the YMCA. Mr. Groves attempted to start a fight with Mr. Foley, and Mr. Foley shot him, killing him instantly. After Groves had been shot, the son jumped on Mr. Kirk and Mr. Kirk shot him in the chest. Groves' son survived and recuperated from his wound. After the shootings, things quieted down in Widen. Early in

1942, the NLRB held its second election, and the UMWA was defeated again, but by only a few votes.

It was the consensus of those who were familiar with the situation in Widen that, if Mr. Groves had not been shot and killed, the UMWA would have won the election.

Foley and Kirk were released under heavy bond pending a preliminary hearing. After the preliminary hearing, they were bound over to the grand jury and later indicted. In the meantime, however, they were released under bond. Mr. T. C. Townsend, the chief counsel for the UMWA and attorney for both Foley and Kirk, had Foley and Kirk go down into eastern Kentucky and join the Masonic Lodge, and he also had them get a Masonic pin to wear on their coat lapels. They were instructed by Mr. Townsend to get closely acquainted with Judge Fisher, who was a Thirty-third Degree Mason; to be frequent visitors of the Masonic Lodge meetings; and to get personally acquainted with those who belonged to the Masonic Lodge. Foley and Kirk became very close friends with Judge Fisher. Mr. Townsend was a Thirty-second Degree Mason himself. Mr. Townsend managed to keep their cases in litigation for about three years and later the cases were dismissed; neither Foley nor Kirk served any penalty for the shootings. However, no further effort was made by the UMWA to organize the Elk River Coal and Lumber Company at Widen.

CHAPTER X

Captive Coal Mine Strike

Early in the fall of 1941, Mr. Bittner had another luncheon for the representatives of District 17 in the old Daniel Boone Hotel in Charleston. After the luncheon, Mr. Bittner related to the officers of District 17 the conversation Mr. Lewis, Phil Murray, Tom Kennedy, and Mr. Bittner had had recently with Ben Fairless, chairman of the Board of United States Steel Corporation, in New York. Of course, Mr. Fairless took the position that the UMW had never shown him that they represented the employees of U.S. Steel's mines. Until such time that the union could show that, he would not negotiate any agreement with the UMWA. The coal mines owned by the steel companies in this country were called "captive coal mines," which meant that the steel industry mined and used its own coal to make steel, and its coal was not used for commercial purposes. Prior to 1941, none of the captive coal mines in this country were under contract with the UMWA. When the issue of the captive coal mines was made a national issue prior to the coal commission named by President Roosevelt to study the matter, President Roosevelt offered to write to all of the employees in the captive coal mines and ask them to join the UMWA. Vice-President Murray was favorable to the President of the United States writing these letters. However, Mr. Lewis opposed it.

Mr. Bittner pointed out that he wanted us to recruit enough pickets from each field and go into Gary Hollow with a very large number of pickets and put pressure on U.S. Steel's Gary operations so that they could not operate their mines. The UMWA had only a handful of employees at U.S. Steel who had indicated their willingness to join the union. So, on the morning of September 21, 1941, we had all of the UMWA pickets from the different areas meet about 2:00 a.m., as we entered Gary. We had approximately thirty-five hundred pickets, and we formed an automobile caravan with these pickets and proceeded to drive to Gary's mine operations with our lights on and blowing our car horns at that hour of the morning. About 4:00 a.m., several of the UMW pickets went to the stable where U.S. Steel housed its mules that they worked in each of their mines. All of the coal hauling at U.S. Steel's mines was done by mule haulage. In other words, they had no mine locomotives; all of the coal haulage inside of the mine was hauled by mules. So, these UMW pickets opened the barn doors and scattered the mules into the mountains so that the company would not be able to work the mines because they had no way of hauling the coal. The employees of U.S. Steel did not report to work that morning, and later in the morning, about eleven o'clock, Mr. Larry O'Toole, who was the general manager of U.S. Steel's Gary operation, came down and asked Mr. William Blizzard, who was vice-president of District 17, and who was in charge of the UMW pickets, to come to his office; that Mr. Harry Moses, the president of U.S. Steel's mine operations in Pittsburgh, wanted to talk to him on the telephone. So, Mr. Blizzard went to Mr. O'Toole's office and talked to Mr. Moses. Mr. Moses proposed to Mr. Blizzard that if Blizzard would withdraw the UMW pickets from Gary, U.S. Steel would not attempt to work its coal mines until something was done about the captive coal mine strike. Blizzard agreed to Moses' request, and when he got back from U.S. Steel's office at Gary, he met with the UMW pickets and told them that the union was withdrawing and to go back to their respective areas. This is what Mr. Blizzard and the UMWA wanted U.S. Steel to do because the UMWA did not have enough support from miners in Gary to even request a NLRB hearing for an election. Actually, the organization of U.S. Steel's employees at Gary was the largest bluff that was ever perpetuated on the employees of the steel industry.

While U.S. Steel lived up to its commitment not to operate the mines, President Roosevelt appointed a coal commission consisting of John R. Steelman, chairman, who was administrative assistant to President Roosevelt, to make a study of the captive coal mine situation and report back to the president as soon as the commission was able to reach a decision in this matter. The other members of the commission were Ben Fairless, chairman of the board of U.S. Steel, and John L. Lewis, president of the UMW. The commission held several closed-door meetings, and the commission made no comments regarding the captive coal mine strike. On December 5, 1941, the commission released its decision. It was a two to one decision requiring the captive coal mines of this country to sign the 1941 UMWA contract, Mr. Fairless dissenting.

On Sunday, December 7, 1941, the Japanese bombed Pearl Harbor, and if the commission had not made its decision two days prior to Pearl Harbor, the captive coal mines would never have been organized, because captive coal mines would never have gone on strike after Pearl Harbor. About the fifteenth of December, the steel industry and the UMWA signed the 1941 agreement covering all of the captive coal mines in the United States and Canada.

CHAPTER XI

The Formation of Southern Coal Producers Association

Shortly after the 1941 Wage Agreement was negotiated, eliminating the forty-cent differential between the northern operators and southern coal operators, James D. Francis, chairman and chief executive officer of Island Creek Coal Company, the largest coal producer in the southern Appalachian region, set out to form a coal association among the companies producing coal in the southern Appalachian area. This was due primarily because John Lewis had forced the southern operators to sign a wage agreement that the northern operators had negotiated, and no negotiations had been conducted by the union with the southern coal companies. This antagonized the southern coal operators, and they were determined to form an organization covering the coalfields of eastern Ohio, southern West Virginia, Kentucky, Tennessee and Alabama. This area represented approximately 200 million tons of coal annually, and they were determined not to be ignored in future wage negotiations.

James Francis was later instrumental in forming the Southern Coal Producers Association (SCPA), covering all of the mines in the southern Appalachian area; on October 17, 1941, a charter was issued by the Secretary of State of West Virginia incorporating the SCPA, a West Virginia corporation.

After the incorporation, a meeting was held in Washington, D.C., and representatives were selected as directors of the SCPA from the Appalachian area. Edwin R. Burke, former U.S. Senator from the state of Nebraska, who was defeated for reelection in the landslide election of President Roosevelt for a third term, was selected as the first president of the association. Walter R. Thurmond, former president of Logan Coal Operators Association during the Blair Mountain Mine War, and a former mine operator in the Logan area, was selected as secretary of the association.

"Walter R. Thurmond was born April 30, 1881 near Oak Hill, Fayette County, West Virginia. The community is now known as Minden. Mr. Thurmond was educated in the public schools of Oak Hill, the Fayetteville Academy and West Virginia University. He graduated from the University in 1905 with a degree of Bachelor of Science."[1] He was a classmate and very close friend of Sen. M. M. Neely of West Virginia.

"From 1905 to 1912, Mr. Thurmond was employed by the New River Company, first as an engineer and then as a mine superintendent. From 1912 to 1933, he operated his own mines in Logan County. In 1933 he accepted an appointment by President Roosevelt as Collector of Internal Revenue for West Virginia. From 1937 to 1941 he served as President of the West Virginia Board of Control. Mr. Thurmond returned to the coal industry in 1941, and helped organize the Southern Coal Producers Association. He was elected secretary of the Association and served in that capacity until his retirement in 1958."[2]

Mr. Thurmond's wide experience in the coal industry; his former employment in an administrative capacity, both in federal and state government; his articulate manner in analyzing coal production statistics; and his reputation as one of the most outstanding individuals in the coal industry, played a part in his selection as SCPA's secretary. Mr. Thurmond was regarded as one of the most influential and knowledgeable of the coal industry's elder statesmen. In 1964, Prof. Robert F. Munn of West Virginia University prevailed upon Mr. Thurmond to write a book entitled, *The Logan Coal Field of West Virginia.* This book can be found in the library of West Virginia University and in most public libraries throughout West Virginia.

The principal office of SCPA was located in Washington, D.C., and the office of the secretary, who handled the financial part of the association, was located in Charleston, West Virginia.

After the announcement of the formation of the SCPA, John L. Lewis vowed that as long as he was president of the UMWA, he would never sit down and negotiate a contract with an anti-union organization. He branded Francis as being the instigator in forming this organization and charged that he was out to destroy the UMWA in the southern coalfields, and had been influential in getting anti-union, ex-Senator Burke of Nebraska as its president.

1. Walter R. Thurmond, *The Logan Coal Field of West Virginia,* p. 13.
2. Ibid., p. 13.

Mr. Lewis totally ignored the leadership of the SCPA and would not have any dealings with Senator Burke.

Sometime in 1948, Senator Burke resigned as president of the SCPA and the directors selected Joseph E. Moody of Connecticut, who had a degree in forestry, as its new president. During the tenure of Mr. Moody as president of the SCPA, he had access to Mr. Lewis in getting information as to the interpretation of the Appalachian Wage Agreement from time to time. Over a period of several years, Mr. Moody and Mr. Lewis cultivated a close friendship. However, Mr. Lewis was still adamant and continued his refusal to negotiate with the SCPA.

In 1959, Mr. Lewis was instrumental in forming the National Coal Policy Conference together with George H. Love, chairman of the Board of Consolidation Coal Company, Pittsburgh, Pennsylvania.

On February 25, 1959, the formation of the National Coal Policy Conference (NCPC) was announced by George H. Love in Washington, D.C.

The NCPC consisted of the mine workers union, coal operators, coal carrying railroads, seven power companies, and mine equipment manufacturers. The primary responsibility of this organization was to lobby for legislation in the Congress of the United States that the members of this organization were interested in.

Due to the friendship that developed between them, Mr. Lewis was successful in having Mr. Moody named as president of the NCPC. This was a fulltime job for Mr. Moody, and he went about organizing his office and staff in Washington, D.C., and resigned as president of the SCPA.

In April of 1960, SCPA named C. W. Davis as its new president. Mr. Davis was a lawyer by profession, and after he had graduated from the University of Tennessee, he became Executive Secretary of the Big Sandy Coal Operators Association in Pikeville, Kentucky, and was later employed by the president of the Bituminous Coal Operators Association (BCOA), Edward G. Fox, as his vice-president. Davis left BCOA in April of 1960 and became president of the SCPA.

During Davis's tenure as president of the SCPA, many of the coal operators who were formally affiliated with SCPA dropped out of the association, and this defection continued, until at last, it reached the point that the remaining companies in SCPA could not continue this organization as a viable organization.

In July of 1964, the office in Washington, D.C., of SCPA and the Charleston, West Virginia, office were closed, and the records of SCPA were transferred to the Kanawha Coal Operators Association in Charleston, West Virginia, to be maintained by that organization until such time as SCPA could be revitalized.

In the meantime, Mr. Davis was employed by Eastern Coal Company which was owned and operated by Lawrence E. Tierney. Mr. Tierney employed Mr. Davis as his general manager, with headquarters in Bluefield, West Virginia. Mr. Davis continued to serve in this capacity for Eastern Coal Company in Stone, Kentucky, and later, after BCOA had accepted Southern Coal into its membership, Mr. Davis was selected as one of the three trustees of the UMW Health and Retirement Fund.

Mr. Davis served in this capacity until his death from cancer in the spring of 1982.

Mr. Lewis lived up to his promise that he would never negotiate any agreement with SCPA.

The SCPA went out of existence with the closing of the Kanawha Coal Operators Association in 1984, and the records of Southern Coal Producers and Kanawha Coal Operators Association were donated to the West Virginia Department of Archives and History.

CHAPTER XII

Ouster of Phillip J. Murray

In the early part of 1942, Mr. Lewis began to plan to remove his very close friend, Phillip J. Murray, as vice-president of the UMWA. In the December 1941 election for International Officers of the UMWA, Phillip J. Murray, for the first time since he had run for international office, received more votes for the vice-presidency than John L. Lewis did for the presidency of the UMWA. This was attributable to the reaction of the mine workers to Mr. Lewis's being opposed to President Roosevelt for a third term. Of course, this did not set well with Mr. Lewis. Since he opposed President Roosevelt for a third term, and President Roosevelt had offered to write letters to each miner in the captive coal industry, and Vice-President Murray had been willing to accept President Roosevelt's offer, these vote results antagonized Mr. Lewis all the more.

In the early part of May 1942, Mr. Lewis summoned all of the international officers, District 17 representatives and officers to meet in Washington on May 25, 1942, together with the International Executive Board. At the beginning of this meeting, he called for a report by the commission that he had appointed earlier in the year to investigate contract violations in District 17. One of the major violations he accused the operators of in District 17 was the tare weight. The tare weight was a method used at each mine location to weigh the impurities from the coal seam that was loaded in a mine car. The company weigh boss and the union checkweighman at the mine would weigh a sample car of coal and then weigh the impurities that were taken out of the coal. The amount of the impurities was then deducted from the total to yield the amount of coal to be credited to the miner. In most instances, the tare weight deduction used would be different at each mine location, and in some instances, the deduction for impurities (tare weight) would range from two hundred pounds per car and run up to as high as one thousand pounds. Whatever tare weight the management and the union agreed to at each particular mine, they would use this as an average tare weight for the whole mine.

Beyond this, the 1934 Coal Wage Agreement carried forward to the 1941 Wage Agreement, provided that a miner had to remove from the coal seam six inches of impurities without any compensation for this extra work. For all over six inches, he was paid per linear foot for the yardage on this work.

When the announcement of the commission to investigate contract violations in District 17 was made to the press in Washington, D.C., Mr. Lewis charged that the operators were taking money out of the pockets of coal miners and building coal processing plants at the expense of this tare weight system at the various operations. Mr. Lewis knew when this statement was issued that it was an erroneous statement, and there was no basis for this charge. The real motive behind this commission to investigate District 17 was because District 17 was the largest district in the jurisdiction of the mine workers. It comprised two-thirds of the coal producing areas in West Virginia, and the membership of District 17 was in excess of 125,000 members. The coalfields covering District 17 were the Kanawha, Logan-Greenbrier, New River, Winding Gulf, Pocahontas and Williamson coalfields in southern West Virginia. Van A. Bittner, a very close ally of Phil Murray, was president of District 17 covering this vast coalfield area. Mr. Lewis was well aware of the potential power that Van Bittner had over this vast coalfield area and that he could not successfully take any action against Bittner or Murray as the situation existed. Mr. Lewis reasoned that the best way to proceed against Bittner and Murray was to use an excuse to send a commission into District 17. The ultimate purpose of this commission was to come up with a plan that would minimize the influence of Bittner and Murray in this area. The commission made several trips to Kanawha, Logan, Mingo and McDowell counties' coalfields, but at no time was any coal operator in this area cited for any contract violations. Of course, this commission was composed of the most trusted lieutenants of Mr. Lewis. They were as follows: Major Percy Tetlow, chairman and international statistician; John Ghizzoni, international board member, District 2, Central Pennsylvania; Charlie Funcannon, international board member, District 11, Indiana; Curtis Mundell, international board member, District 12, Illinois; Frank

D. Wilson, international board member, District 14, Pittsburgh, Kansas.

When the commission made its final report at a meeting on May 25, 1942, to the International Executive Board and the staff who were present for the UMWA, the commission complimented the officers and members of District 17 for the outstanding service they were providing the membership. They also complimented the officers and staff of the district for their capabilities in handling the affairs of the union. However, Major Tetlow, in his report, pointed out that District 17 was too large, and the area too spread out, and it was hard to give the membership the service that it expected. Also, it was too far for some of the miners to travel to Charleston, West Virginia, from the different areas to discuss their problems. The commission recommended to the Executive Board that District 17 be split in two and recommended that the International Union reestablish District 29 with headquarters in Beckley, West Virginia. District 29 would have jurisdiction in the Greenbrier, New River, Pocahontas and Winding Gulf coalfields, formerly part of District 17. The International Executive Board adopted the recommendation of the commission and the commission was dismissed.

Around the first of June, it was announced by President Lewis that District 29 was being reestablished and that the secretary-treasurer of District 17, George J. Titler, would become president of District 29, that Nick Allio would become vice-president, and Durf Stamper would be the secretary-treasurer. In the same announcement, Mr. Lewis announced that Major Tetlow would become the acting president of District 17, William Blizzard would remain as vice-president, and Paul K. Reed of Illinois would succeed George J. Titler as secretary-treasurer.

In the meantime, this solidified separate Districts 17 and 29 the way Mr. Lewis planned it because the new officers were handpicked, and he had both districts set up so that both districts would carry out his policies. He felt safe that there would be no autonomy movements attempted in either of these districts.

After the International Executive Board approved the action of the commission, in the next order of business Mr. Lewis charged that some of the officers and members of the CIO had criticized and vilified members of his family for his action in opposing President Roosevelt for reelection and that the officers of the CIO did nothing to correct the situation. These charges and countercharges were brought out into the open in this meeting, and on one occasion, John Lewis accused Phil Murray, vice-president of the UMWA, and then the president of the CIO, that Murray had compared the Lewis family to a "Jap" in a press conference in New York City. Vice-President Murray leaped to his feet and hit Lewis on the left shoulder with the palm of his hand and uttered "You know that is a Goddamn lie, John," and denied this accusation. Mr. Lewis did not respond to what Phil Murray had just said. The meeting continued through Tuesday and Wednesday and in these sessions, Phil Murray denied all of these accusations.

On Thursday morning, May 28, about 12:30 p.m., Phil Murray got up and requested that Mr. Lewis excuse him for the afternoon because he had an appointment with President Roosevelt and the members of the War Labor Board at 1:00 p.m. to discuss matters that affected labor in general during the war period. Phil Murray hesitated for about five minutes, and John Lewis did not respond whether he was excusing him or not. Murray then went over to where David McDonald was seated and picked up his coat and hat. David McDonald also left with Murray. McDonald was secretary to Phil Murray at this particular time and they left, going to the White House. Immediately after Murray's departure, Mr. Lewis convened the International Executive Board, and in the presence of all of those who were in attendance at this meeting, he charged that Phil Murray had accepted a position as president of the United Steelworkers Union and could no longer serve as a vice-president of the UMW. He then declared Phil Murray's office vacant. John Ghizzoni, International Executive Board member from District 2, moved that Mr. Lewis's point was well taken and moved declaring the office of Vice-President Murray vacant. International Executive Board member Robert Livett of District 18 seconded Mr. Ghizzoni's motion. When Mr. Lewis called for the question on the motion, Martin Wagner, who was a member of the International Executive Board and president of District 50, rose to speak. He pointed out to the international board members that he had never seen such conduct (such as that which had been used in vilifying Phil Murray) carried on by people who confessed to be union people. He also pointed out to Mr. Lewis that Lewis was using the same tactics to get rid of someone as the coal operators used to get rid of someone with a lot of untruths and falsehoods, and that Phil Murray was an honorable man and should not be treated in this manner by having his office declared vacant without him being present to defend himself. Wagner continued to voice his objection and said he wanted to be recorded as being opposed to the motion. Mr. Lewis then proceeded with the motion, and the motion was carried with the dissenting vote cast by Martin Wagner.

After Mr. Wagner cast his vote, he announced he was resigning as International Executive Board member and District President of District 50 immediately and stated he would have no part of an organization that conducted its business in this manner. Of course, those who were in attendance did not expect Martin Wagner to get out of there without getting beat up because there had been several undesirables from the Chicago Southside imported for this meeting to manhandle anybody who got out of line. However, this did not happen, and Wagner left the meeting.

Then, another motion was made by John Ghizzoni of District 2 and seconded by Robert Livett of District 18, Calgary, Canada, that John O'Leary, a member of the International Executive Board from District 5, Pittsburgh, Pennsylvania, be designated as the new vice-president of the UMW. Mr. Lewis put through the motion, and the motion carried unanimously.

After both of these motions were out of the way, Mr. Lewis, addressing those persons who were in attendance who had witnessed the ouster of Phil Murray, stated that he believed in an old adage his father handed down to him many years ago. He then quoted his father and the quote was, "Son, if you eat my bread, you'll sing my song." As soon as Mr. Lewis uttered these words, the audience responded by jumping to their feet and applauding him because they all understood what Lewis meant and that was, if you were going to stay on the UMWA payroll, you were going to sing John L. Lewis's song. Prior to adjourning the meeting, Mr. Lewis asked Mr. Bittner to stand up, and he then requested whether or not Mr. Bittner was going to be in town the next morning, because he wanted to see him in his office. Mr. Bittner answered that he would see him. Later in the evening, Mr. Bittner requested the officers and representatives who were in attendance from District 17 meet him down at the Carlton Hotel, now known as the Sheraton-Carlton, where Phil Murray, Van Bittner and Tom Kennedy were all staying. Mr. Murray was very bitter about the manner in which he had been removed as vice-president of the UMW and so expressed himself to us at this meeting. Mr. Bittner advised the representatives of District 17 that he would be in Charleston on Monday and would have a meeting with all of us at that time.

In this meeting in Charleston, Mr. Bittner was very critical of Mr. Lewis and critical of the manner in which Mr. Lewis had handled this affair. He told us that on Friday morning when he met with Mr. Lewis to discuss some issues in District 17 and after he had left his office and was returning to Charleston, Mr. Lewis had a press conference and stated that he had accepted Bittner's resignation as president of District 17. Mr. Bittner told us that his resignation was not even discussed in the meeting in Washington, and that he did not resign, and he charged John L. Lewis with being a most deceitful person. He admonished all of us to continue to do our work in carrying out the instructions of the newly announced officers of District 17, and in the event any of the representatives would lose their jobs as a result of the new leadership in District 17, he assured those present that he would hire them with the CIO and pay them their salary and they would not have to leave West Virginia to work. This was Mr. Bittner's final action as president of District 17.

CHAPTER XIII

1943 Wage Agreement

In the first part of October 1942, the UMW held its constitutional convention in Cincinnati, Ohio, at the Netherland Plaza Hotel as the convention headquarters. During one of the sessions of the convention, the convention adopted the officers report recommending that the UMWA withdraw its affiliation with the CIO and all of the UMW representatives who held office in the CIO, withdraw immediately from the CIO. This notification was forwarded to the CIO at its convention in Detroit, Michigan, in November 1942.

As a result of this action, the UMW became an independent organization, with no affiliation in the CIO or the AFL.

In March 1943, the UMW began negotiations with the coal operators of the Appalachian area for a new contract to replace the one that was due to expire April 1, 1943. Negotiations were short-lived because the coal operators refused to negotiate a new agreement. "By March 22, when no progress had been made, President Roosevelt asked that negotiations be continued beyond the time of contract expiration."[1] The UMW agreed to an extension until April 30. However, the operators attitude toward negotiations remained unchanged. "Their secret purpose was soon revealed, for on April 22 Madam Perkins . . . (Secretary of Labor) . . . certified the case to the NLRB."[2]

However, the UMW refused to go along with this, and on April 30 the miners went on strike. On May 1, 1943, President Roosevelt issued an executive order placing the coal mines

1. International Executive Board, United Mine Workers of America, *John L. Lewis and the International Union, United Mine Workers of America: The Story From 1917 to 1952,* p. 147.

2. International Executive Board, United Mine Workers of America, *John L. Lewis and the International Union, United Mine Workers of America: The Story From 1917 to 1952,* p. 149.

under the jurisdiction of the United States Government and designated the Secretary of the Interior, Harold Ickes, who was also Solid Fuels Administrator, as the operating manager of all coal mines in the United States. Secretary Ickes met with Mr. Lewis on May 2, 1943, and requested that the mine workers resume operation immediately. "Nevertheless, President Roosevelt went on the radio that very night and, in utter disregard of our action, appealed to the coal miners to get back on the job."[3] He also stated that the American flag would be flown over all coal mine tipples in the nation, and he expected the miners to resume work the next morning. After President Roosevelt's broadcast to the nation's miners, the miners did not resume work, and on May 17, 1943, a conference was held between Mr. Lewis and Secretary Ickes, and Mr. Lewis agreed to another 30-day extension and the miners resumed operation.

The miners dispute was in and out of the War Labor Board, and the miners union was unable to accomplish much through the WLB. The mines continued under government operation through November 3, 1943, when Mr. Lewis and Secretary Ickes reached an agreement known as the Lewis-Ickes Agreement.

In May of 1943, the U.S. Supreme Court ruled that miners were entitled to portal-to-portal pay, and portal-to-portal pay was written into the Lewis-Ickes Agreement which provided a work day of nine hours from portal-to-portal because of the war effort and forty-five minutes was travel time at time and one half. It became the first time in this country that any major industry was required to pay portal-to-portal pay or travel time pay. Prior to this agreement, miners were required throughout the industry to travel to their working place on their own time. In other words, a miner's time started at his working place and ended at the working place, and there was no provision for travel time pay. This was one of the major settlements achieved by the UMWA and was ruled on by the U.S. Supreme Court that the miner was entitled to travel time pay.

Other concessions were made to Mr. Lewis by Secretary Ickes. One was an increase in the miners vacation bonus from $20.00 to $50.00, and they waived the vacation period due to the national emergency. Also, the employer was required to provide tools to the miners and pay for their work clothing, and wages were increased $1.50 per day. This agreement covered the period from November 3, 1943, to March 31, 1945.

3. Ibid., p. 149.

In addition to the above concessions, it was agreed that all employees, both outside and underground miners, would be paid a bonus of forty dollars each in settlement of all back liability of portal-to-portal pay.

On December 17, 1943, the operators of the Appalachian area agreed to the Lewis-Ickes Agreement and signed it. As a result of the operators acceptance of this agreement, the government relinquished its operation of the coal mines.

President Lewis, between 1943 and 1944, atttempted to unite the labor movement by extending an invitation to Phillip Murray, president of the CIO and William Green, president of the AFL, for a joint conference to discuss the unification of both of these labor organizations. Both Mr. Murray and Mr. Green ignored Mr. Lewis's peace efforts, and this left John Lewis outside the labor movement, with no influence within organized labor. This was mostly brought about because of Lewis's refusal to go along with and support President Roosevelt's bid for a third term.

The fact that Phil Murray and William Green, both former officers of the UMW, did not respond to Mr. Lewis's letters, was a great disappointment to Lewis because he wanted very badly for the labor movement to be united. Both the AFL and CIO isolated Mr. Lewis from their organizations.

CHAPTER XIV

Establishment of Welfare and Retirement Fund

"The 1945 bituminous negotiations officially began when the UMWA, as required by the Smith-Connally Act, notified the Secretary of Labor that a "labor-management dispute" existed. This provision of the law was particularly obnoxious to the miners because it sometimes required unions to certify a dispute that didn't exist. In the miners' case in 1945, it merely meant that the industry's wage agreement expired in thirty days.

"Secretary of Labor Perkins offered to mediate the differences and the War Labor Board tried to get into the act, but the contract was arrived at by collective bargaining."[1]

"The new agreement, signed April 11, was for one year instead of the usual two. Both parties recognized that rapidly changing economic conditions made more frequent negotiations desirable. The basic hourly wage rates were retained but a new overtime arrangement increased the pay of the $7-a-day inside men from $57.06 to $64.50 a week. A long delayed equalization of rates in unit mechanized production gave drillers and shooters a larger share of the benefits of mechanization.

"The real battle for the United Mine Workers of America Welfare and Retirement Fund began on March 12, 1946, the date the bituminous contract negotiations began. The miners had resolved to win that battle no matter what it cost them in time or money.

"In 1946, conditions seemed much more favorable for a winning fight than they had been in the spring of 1945 when the fund had been proposed and then dropped in the interests of a quick settlement.

1. International Executive Board, United Mine Workers of America, *John L. Lewis and the International Union, United Mine Workers of America: The Story From 1917 to 1952,* pp. 157-58.

"The war was over. Veterans were returning to civilian life. There was a new president in the White House. Franklin Delano Roosevelt had died in Warm Springs, Georgia, on April 12, 1945, just prior to the end of the war in Europe. His successor, Harry S. Truman, an apparently amiable, humble man, ex-Senator from Missouri, seemed to be more friendly to labor than Roosevelt had been during the war.

"Other omens, however, were not so auspicious. Commodity prices were skyrocketing. The OPA seemed powerless to control food prices. The government still had a law on its books which the miners called the "Smith-Connally Slave Statute."[2] This act was wartime legislation and only applied in the event of a war.

"Conferences began March 12. After preliminary discussions lasting three days, the miners presented their proposals to the operators on March 15."[3]

"The UMWA then presented the results of a survey of medical conditions in mining areas which had been conducted by a miners' committee, headed by Harrison Combs, UMWA associate counsel. The survey showed the degraded condition of sanitation and medical facilities in mining towns."[4]

On March 31, 1946 the National Coal Wage Agreement expired, and the coal miners were back out on a nationwide strike. The government tried to keep the mines in operation, but to no avail. When March 31 rolled around, the miners refused to work because they did not have a contract. The U.S. Government urged the miners to resume operation, and the miners ignored the government's plea, so President Truman invoked the Taft-Hartley law and ordered the miners back to work, and the government took the mines back over.

"The week beginning May 22 and ending May 29 was one of the most hectic weeks in the history of coal controversies, as well as government labor policies. The government seized the coal mines on May 22. The railroad strike immediately followed. On May 25, President Truman proposed his ill-advised legislation to take over industry and draft manpower and force men to work under the strong arm of the Army and Navy. Then came the capitulation of the engineers and trainmen and the calling off of the railroad strike.

2. International Executive Board, United Mine Workers of America, *John L. Lewis and the International Union, United Mine Workers of America: The Story From 1917 to 1952*, p. 158.

3. Ibid., p. 159.

4. Ibid., p. 161.

"While all this was in progress, President Lewis, Vice President John O'Leary, Secretary-Treasurer Thomas Kennedy, and John Owens, president of District 6, were conferring, first at the Department of the Interior and then at the White House. Almost daily, conferences were being held with Secretary of the Interior J. A. Krug and his deputy, Admiral Ben Moreell. At the White House, President Truman, John Steelman and John Snyder participated.

"The climax came when a contract, the terms of which were agreed upon the night before, was signed by Secretary of the Interior Krug for the government and John L. Lewis for the UMWA in the President's office. The coal strike was over. The movie and news photographers were lined up in rows of five deep to photograph the principals, as Secretary Krug, accompanied by Admiral Moreell, and President Lewis, with Mr. O'Leary and Mr. Owens, president of District 6, emerged from the White House office."[5]

In August of 1946, a pension trust was established providing a one hundred dollars per month pension for coal miners who retired and had worked in the mines twenty years of the last thirty at the time of their retirement. The fund also provided medical and hospitalization coverage and sickness and disability payments. The fund was operated by three trustees: John L. Lewis, chairman; Senator Styles Bridges of New Hampshire, who represented the public; and H. G. Schmidt, chairman of the board of North American Coal Company, Cleveland, Ohio, representing the coal industry. The benefits were to be financed out of trust funds financed by the coal operators on the basis of five cents per ton.

The greatest achievements in the UMWA, particularly under the leadership of John L. Lewis, came in the period 1940 through 1950. After 1948, he embarked upon making the UMWA the greatest labor union in the world so that it could not be destroyed from within or without. And some of his significant accomplishments were:

*The elimination of the wage differential between the miners who worked in the mines South of the Ohio River who received forty cents per day less than the miners in the North;

*The shift differential;

5. International Executive Board, United Mine Workers of America, *John L. Lewis and the International Union, United Mine Workers of America: The Story From 1917 to 1952,* p. 165.

*The vacation bonus and the designated holidays to be observed;

*Time and a half for overtime;

*Portal-to-portal pay;

*Furnishing of miner workers' clothing and free working tools for miners;

*The establishment of the Welfare and Retirement Fund covering coal miners who were employed in the mining industry, and;

*The building, for the first time, a chain of ten miners hospitals throughout the Appalachian area from Middlesboro, Kentucky in the south to Beckley, West Virginia in the north, by the United Mine Workers

This was the first time that a labor union in this country had undertaken building such a vast chain of hospitals to service its membership and their families in this area. The hospitals were completed in the early 1950s. The union continued to operate these hospitals for several years; however, since the union did not have any prior experience in operating hospitals, the hospitals soon became very costly. The union prevailed upon the Presbyterian church to take over the hospitals and operate them on a nonprofit basis. Later, the church formed a corporation known as the Appalachian Regional Hospital Incorporation, and the hospitals have been in operation on a nonprofit basis up to the present. In addition to these accomplishments during this decade, the basic miners wages were raised from the hourly rate $.88 and a day rate of $6.16 in 1940 to $1.87 hourly rate and $14.94 day rate in 1950.

The coal contract signed on May 29, 1946, was under the government operation of the mines when President Truman invoked the Taft-Hartley law and designated the Department of the Interior, Julius A. Krug, Secretary of the Interior, as the operating manager, he then designated Admiral Ben Moreell of the U.S. Navy as the operating manager for the U.S. Government. The Krug-Lewis Agreement signed on May 29, 1946, provided for a Welfare and Retirement Fund to cover the employees of the coal industry. The fund provided a payment by the industry of five cents per ton for each ton of coal produced by the operator. The coal operators initially refused to have anything to do with this agreement. The Krug-Lewis Agreement continued with the mines under government operation until July 7, 1947 when the bituminous coal operators finally signed the Krug-Lewis Agreement and the government relinquished its operation of the coal mines and turned them back over to the coal operators.

The Krug-Lewis Agreement also provided a mine safety program; coverage under Workers Compensation, occupational disease laws; and a survey of medical and sanitation facilities. All of these issues had been raised by the mine workers union and were covered in the Krug-Lewis Agreement. After the operators signed this agreement, President Truman appointed Admiral Joel Boone, retired, U.S. Navy, to make a survey of the health and sanitation conditions that existed throughout the coal industry. Specialists named in addition to Admiral Boone were to study housing, health and sanitation.

This commission traveled throughout the bituminous coalfields in the United States, and when it made its final report public, the report cited some of the worst conditions that prevailed in the entire industry. In the major portion of the industry there was no water and bath facilities in the homes of the miners. The only facility available for water was a spigot located about twenty yards from the miners homes. Generally, there were approximately six homes to each spigot. These spigots were outdoors, and therefore, the miners had to carry their water home to take a bath and to use it for cooking their food (there were no bathhouses located in most of the coalfields in this country).

When the Admiral Boone Commission made its final report, it was highly critical of the conditions that existed in the coalfields. In the meantime, the coal companies who owned these mining towns were fearful that the government, as a result of Admiral Boone's survey, would compel the coal companies to improve the housing and sanitation facilities of the miners, so the coal companies throughout the industry began to sell company houses to their employees. The miners purchased their homes; renovated them; installed running water and bathing facilities; and made additional improvements to their homes.

CHAPTER XV

Miners Purchase National Bank of Washington

Between 1946 and June 25, 1948, a new coal agreement was made, and during this period, things began to improve in the health and sanitation facilities in the coal industry. The agreement signed on June 25, 1948 continued through June 30, 1949. Mr. Lewis was determined to build the UMW union into the most outstanding labor organization in this country. He had given up trying to promote unity in the labor movement because both labor leaders in the AFL and CIO totally ignored Mr. Lewis's advice, and it became apparent that he had no influence outside of his own union.

Due to the creation of the Welfare and Retirement Fund which required the coal producers to pay into the fund five cents per ton on each ton of coal from May 29, 1946, through the early part of 1949, this amounted to millions of dollars that accumulated in the fund, and these monies were deposited in the National Bank of Washington. Mr. Lewis requested of the bank that the UMW be given some voice in the policymaking decisions of the bank because of its large accumulation of welfare funds on deposit in the bank. The directors of the National Bank of Washington refused to consider Mr. Lewis's request, and by the middle of 1949 ". . . the United Mine Workers of America bought controlling interest in the National Bank of Washington . . . because the union could not depend on other banks to fund its battles against big business. The UMW was fighting the Southern Coal Producers Association back then and taking on all other comers in the struggle for better wages and benefits for the people who dug and hauled the nation's coal."[1]

1. The *Washington Post*, March 6, 1985.

The UMW purchased seventy-six percent of the stock of the National Bank of Washington, and the cost of this stock was approximately $25,721,643, giving the union complete control of the oldest and also one of the largest banks in the city of Washington.

During Mr. Lewis's tenure as president of the union, he did not permit the UMW organization to interfere or use undue influence in the policymaking decisions of the bank. The directors and president of the bank continued to operate and build the assets of the bank in a businesslike manner.

Later in 1950, Mr. Lewis purchased the lot adjacent to the UMWA building. Also, the UMWA purchased a five-story building on the corner of 15th and K streets where the offices of the *United Mine Workers Journal*, the Contract Enforcement Division and the Mine Safety Department were moved to. These properties were acquired at a bargain price which gave the UMW additional assets, making the UMW organization one of the most powerful labor organizations in this country, both in finance and prestige.

CHAPTER XVI

The Formation of BCOA

After the completion of the purchase of the outstanding stock of the National Bank of Washington, Mr. Lewis continued to strengthen his organization. He proceeded to influence the steel industry to form an organization of its own to negotiate future contracts with the UMWA. In late 1949 and early 1950, Mr. Lewis was able to convince the steel owners to form the Bituminous Coal Operators Association (BCOA). Mr. Lewis enjoyed a very close friendship with Harry Moses, who was then in charge of the coal division of United States Steel (USS). This acquaintance and relationship began in the days when Mr. Lewis worked in the coal mines in southern Illinois with Tom Moses, the father of Harry. Since then, Harry Moses has had a long tenure of employment with USS. Prior to moving to Pittsburgh to be in charge of all USS's mines, he had been manager of USS operations at Gary, West Virginia.

So, the steel industry put up a major portion of the money together with the coal companies that the steel industry purchased coal from, which amounted to several hundred thousand dollars, for the purpose of setting up a pension fund and other fringe benefits to replace the benefits that Harry Moses had earned in the employment of USS. He would not take the job as president of the BCOA unless it provided him with the same commitments he had earned with USS.

The BCOA was chartered July 1, 1950, with Harry Moses as its first president. The first meeting was held in Washington on July 11, 1950, completing the organization of the BCOA.

With the acceptance of the new agreement on March 5, 1950, the increase in the royalty payments from twenty cents per ton in 1948 to thirty cents per ton in 1950 into the UMW Welfare and Retirement Fund, and this vast amount of money accumulating and being on deposit in the National Bank of

Washington, not only did the bank gain added significance throughout the financial markets, this also increased the influence and prestige of the UMWA with many powerful groups in this country. Mr. Lewis rapidly became close friends with some of the financial giants in this country. He became a very close personal friend of Cyrus Eaton, who was both a financial and industrial giant. Mr. Eaton was chairman of the board of directors of Cleveland Trust Company and chairman of the Chesapeake and Ohio Railway. Due to the close friendship between Mr. Lewis and Mr. Eaton, Mr. Lewis was able to convince Mr. Eaton that with the financial support of the UMW, he should consider the purchase of the largest nonunion coal companies in western Kentucky, West Kentucky Coal Company and the subsidiary, Nashville Coal Company. These companies produced approximately 4,160,229 tons of coal per year, and based on the increase to forty cents per ton in royalty payments to the UMWA Welfare and Retirement Fund, this would add approximately $12 million per year to the UMW Welfare and Retirement Fund. In addition, the companies employed approximately two thousand people, and Mr. Lewis foresaw the potential advantages this could have to both the UMWA and the National Bank of Washington.

In 1951 Cyrus Eaton, with the help of the UMW, purchased the entire stock of the West Kentucky Coal Company, and upon completion of the sale, Mr. Eaton sat down with the president of District 23 of the UMWA in Madisonville, Kentucky and signed the National Coal Wage Agreement of March 1950, ushering in a new era in the West Kentucky coalfields.

There was a lot of criticism by the people of west Kentucky against Mr. Eaton for forcing the employees of his companies into the union and into signing the 1950 Wage Agreement. Mr. Eaton continued to operate these companies until sometime in the early part of 1961 when he sold all of his interests to Island Creek Coal Company. In December of 1961, the UMW sold its fifty thousand shares in West Kentucky Coal Company and Island Creek Coal Company became the exclusive owner. So one can readily see the skill and maneuverability that Mr. Lewis possessed in this outstanding transaction in bringing into the fold of the UMW the last large nonunion operation, making a total of ninety-seven percent of the coal industry under contract with the UMWA.

CHAPTER XVII

Lewis Visits Logan

On Saturday night April 21, 1951 about 10 P.M., I received a telephone call from John L. Lewis from the Pikeville Hotel, Pikeville, Kentucky. Mr. Lewis stated that he was on his way back to Washington from Florida, where he had just spent the winter; that he and his daughter, Kathryn, were spending the night in Pikeville and would be following Route 119 from Pikeville to Logan; and since he had never been in Logan, he wanted to know where he could meet me to talk on Sunday morning, April 22, 1951, about 11 A.M.

I advised him that Route 119 came right by the local bus terminal; that next to the terminal was an Ashland Oil filling station; that he should pull into that service station where I would meet him. On Sunday morning, April 22, at precisely 11 A.M., Mr. Lewis pulled into the service station driving his Cadillac automobile with his daughter Kathryn alongside in the front seat. Mr. Lewis got out of the car and asked the service station attendant to service his car, and Mr. Lewis took me back into the grease room where we could talk. His first conversation was about coming down through Stirrat, Barnabus, and Omar where he had stopped to shake hands with most of the colored people who were just coming out of church. Although he had an awful time getting away from them, he enjoyed his short visit. He asked me who owned the Stirrat operation, and I told him it was part of the West Virginia Coal and Coke operation, and their headquarters were in Omar. He asked me if I had met James L. Hamilton, the newly appointed vice-president of the Island Creek Coal Company in Holden. I told Mr. Lewis that I had not yet had an opportunity to meet Mr. Hamilton, since he had been in Holden only about two weeks. Mr. Lewis responded and told me that Mr. Hamilton was the former head of Republic Steel's coal mines, with headquarters in Uniontown, Pennsylvania, and he was responsible for having Mr. Hamilton employed as executive vice-president of Island Creek Coal Company. Mr. Lewis told me that he was a close personal friend of Cyrus Eaton, who was chairman of the board of Island Creek Coal Company, chairman of the board of Cleveland Trust Company, and chairman of the board of the C.&O. Railway. Also, Island Creek Coal Company had formerly had a policy of promoting its own employees to the top management of the company, but Hamilton was the first outsider to move into a top position in Island Creek Coal Company.

He directed me to go to Holden on Monday morning and get acquainted with Mr. Hamilton and offer him any assistance I could. Mr. Lewis was very critical of J. D. Francis, former president of Island Creek, and he told me that he was also responsible for forcing Mr. Francis to retire as chairman of the board of Island Creek. He stated that Mr. Francis was an enemy of the UMW and had organized Southern Coal Producers Association (SCPA) for the explicit purpose of destroying the UMWA. Mr. Lewis was convinced that Mr. Hamilton was in for a lot of trouble because he had been hired over others who had been with Island Creek for many years. Mr. Lewis wanted me to tell Mr. Hamilton that he regretted that he was unable to see him while on his stop in Logan and requested that I convey to Mr. Hamilton that on his next trip to Washington, Mr. Lewis wanted for him to arrange to see Mr. Lewis. I assured Mr. Lewis that I would go to Holden and convey his message to Mr. Hamilton.

In continuing our conversation, Mr. Lewis told me that he was responsible for getting Mr. Eaton to go to west Kentucky and purchase the West Kentucky coal companies and that the union had put up the money for Mr. Eaton to make this purchase.

In servicing Mr. Lewis's custom-built Cadillac, the attendant noticed "John Llewellyn Lewis" inscribed across the motor in big red letters. In the meantime, word got around in Logan that Mr. Lewis was in town, and a crowd of approximately two hundred people began to gather around the service station. Mr. Lewis advised me that the news reporters would be contacting me to find out what Mr. Lewis's business had been and where he was in Logan. He advised me not to give them any information at all; to tell them only that he had been passing through Logan on his way to Washington, and to keep our conversation confidential. It is not mere coincidence that when Island Creek Coal Company purchased West Kentucky Coal Company and Nashville Coal Company from Mr. Eaton, James L. Hamilton was the one that completed the deal with Mr. Eaton, and Mr. Eaton was the one that completed the deal with Mr. Eaton, and Mr. Eaton was the one that forced Island Creek to employ Mr. Hamilton in 1951.

CHAPTER XVIII

The Holden Tragedy

In the early part of 1951, Mr. Lewis received a complaint from Local Union No. 5853 of Island Creek Coal Company, Holden, West Virginia, regarding a dispute involving an individual, Elmer Farley, who was a member of this local union and also an employee and union member at the Island Creek Coal Company mine repair shop. The repair shop came under the jurisdiction of this local union.

Mr. Farley was an industrious individual. He was always doing other work besides his regular work at the mine shop. He rigged up his truck so that on the days that the mines were idle he could haul house coal at a very low cost to the miners, since the miners had Mr. Farley haul house coal for far less than the National Wage Agreement provided for unionized truck drivers. He would haul a two-ton load of coal for one dollar per load. Under the wage agreement, the cost of hauling coal would amount to between three and four dollars per ton that a miner would have to pay if he required the company to haul his coal with union drivers.

The officers of the local union attempted to stop Mr. Farley from hauling this coal on Saturdays and idle days. The local union issued a written notice to Mr. Farley to appear at the next local meeting. Mr. Farley appeared at the meeting. He was told by Charlie Wells, the president, and Major Thompson, chairman of the mine committee that the next time they caught him hauling house coal to the miners on idle days and Saturdays, they would "kick his ass" and inflict bodily harm upon him; that his job was at the mine shop and not hauling coal for the miners.

As a result of this particular issue and the fact that it had been reported to him, Mr. Lewis designated a commission composed of Ray Tomasson, international board member from District 29, Beckley, West Virginia, and Tom Raney, international board member of District 30, Pikeville, Kentucky, to investigate this complaint, and report back to him after their investigation. Mr. Tomasson and Mr. Raney notified the president and the mine committee that they would meet them in the Logan office of the UMW on Friday evening, June 1, 1951 at 7 P.M. to investigate this complaint. The meeting was held as scheduled, and those present including myself were two of my field workers, Nelson Beatty and Jerry Stidham, along with Ray Tomasson and Tom Raney. The representatives of the local union from No. 1 Holden were also present. They were Charles Wells, president; Major Thompson, chairman of the mine committee; Robert Worley, mine committeeman and chairman of the safety committee, and Napoleon Gilbert, financial secretary of the local union.

After listening to the committee's complaint concerning Elmer Farley, Mr. Tomasson told the officers of the local union to personally see that Elmer Farley did not haul any more house coal with his truck, and for them to put an end to his hauling house coal any way they could that would be the easiest and quickest. He encouraged them to use force if necessary. After the committee left the meeting, I engaged Mr. Tomasson in a heated conversation over this matter. I pointed out to him that the instructions he had just given to the committee were not proper, and he should have instructed the officers to proceed under the grievance procedure of the National Wage Agreement and should have had the committee file a grievance with the management so that both parties could sit down and try to resolve the problem. Mr. Tomasson did not appreciate my comments and made it clear to me and the two other field representatives that he and Mr. Raney were the two representatives sent in as the commission to investigate this matter and that it was none of our business. He cautioned us not to try to interfere with the commission's investigation. Both Mr. Tomasson and Mr. Raney were staying at the Aracoma Hotel, and after this argument between myself and Mr. Tomasson, the meeting broke up with no understanding being reached on this matter.

On Saturday morning, June 2, 1951, Elmer Farley drove his truck to the No. 7 Holden tipple to load so he could deliver house coal to a miner at Monaville, West Virginia, which was approximately twelve or fourteen miles from the No. 7 tipple. In the meantime, Major Thompson found out that Elmer Farley had gotten a load of coal at the tipple and was going to deliver the coal at Monaville. He picked up Napoleon Gilbert, a black man, at his home and then drove to No. 21 Holden to get

Charlie Wells, the president of the local union. They then drove to Robert Worley's home and picked him up and proceeded to Lindley Camp at Monaville where Elmer Farley was unloading the coal. When Mr. Farley saw the officers of the local union pull up in Major Thompson's car, he jumped up, ran and got into the cab of his truck. Mr. Farley had a loaded pistol laying on the seat of his truck. He picked up the gun, and when Charlie Wells started getting in his truck from the driver's side to get him, Mr. Farley shot him and killed him instantly. In the meantime, Robert Worley had gone around to the other side of the truck to get in. Mr. Farley also shot and killed him instantly. Then, Mr. Farley got out of the truck, and when Major Thompson saw that he had killed both fellows, he started to run. Farley shot him in the back, and the bullet went clear through him and severed his spine. Napoleon Gilbert never got out of the car. That was the only reason Elmer Farley didn't shoot him. Later, Thompson was taken to a hospital in Huntington and remained in serious condition for several days. After Thompson was released from the hospital, he lived about a year and a half and died as a result of his gunshot wound.

As the chief representative of the UMW in the Logan field, I had participated in the meeting between Tomasson and Raney and the local officers in my office on Friday evening, June 1, 1951. Farley shot these individuals around 10 A.M. on Saturday morning. When the funeral home returned from Monaville with the bodies of Wells and Worley, they called me at home and told me of the shootings. I called William Blizzard, president of District 17, UMWA, in Charleston, West Virginia, at home and relayed to him what had happened. Blizzard instructed me not to get involved in this affair and to be sure not to get the International Union or the district representatives involved in any manner as a result of these shootings. He also instructed me to contact Tomasson and Raney and tell them exactly what had happened. I went to the Aracoma Hotel to see if Mr. Tomasson and Mr. Raney were there, and the hotel clerk told me that they had checked out of the hotel about thirty minutes earlier.

After leaving the hotel, I met two coal miners in front of the hotel, and they informed me that they had relayed to both Tomasson and Raney the news of the officers of No. 1 Holden being shot by Elmer Farley, and that Tomasson and Raney, after finding out what happened, checked out of the hotel and left Logan. Neither one of them ever returned to Logan.

In the meantime, Mr. Farley gave himself up and was placed

in the Logan County jail. Wind of these shootings spread throughout Logan County. The news media, through United Press International and Associated Press, picked up the news of these shootings and it was carried throughout this country. It was widely discussed among the coal miners in Logan County that they should meet in force in Logan and take Farley into custody and hang him. As the rumors spread, the sheriff and the law enforcement agencies became quite concerned over this alleged threat. Meanwhile, on Saturday evening, the sheriff and a couple of his deputies removed Farley from the jail and placed him in the Cabell County jail. Farley, until he was indicted by the grand jury in Logan County, was lodged in the Cabell County jail for about two months. He was then returned to the Logan County jail to await his trial.

At his trial, Farleys' defense was that he committed the shootings in self-defense. In fact, he had all of the members of Local Union No. 5853 subpoenaed to testify in his behalf. All those who testified admitted that Farley had been requested to appear before the local union and that Wells and Thompson had threatened him with bodily injury if they ever caught him hauling house coal again. Based on this testimony, the jury returned a verdict of "not guilty." Farley was released and continued to work in the repair shop until about 1983 or 1984, at which time he retired and began receiving his miner's pension and Social Security benefits. Until his death in 1986, he resided in the house he had always lived in at Holden.

What is most significant about this case is that during the heat of emotion at the time of the shooting of these individuals, most of the miners were going to get revenge for the killing of their brothers; yet, three months later, while Farley was being tried, most of these same people testified in his behalf having forgotten their dead brethren.

The families of the deceased engaged W. E. Flannery, a local attorney, to assist the prosecuting attorney in trying Elmer Farley for first-degree murder. Just before the case was to be tried, I was ordered into Charleston to the district office where I met with president of District 17, William Blizzard, and Secretary-Treasurer R. O. Lewis. I was instructed to contact attorney Flannery upon my return to Logan to advise him not to charge the families of the deceased for his legal services in prosecuting Elmer Farley but to submit his bill to the secretary-treasurer's office in Charleston, and they would pay for his services. Mr. Flannery rendered his bill for fifteen hundred dollars for his legal services in this case, and it was paid by District 17.

About eight months later, I was called back to Charleston and met with William Blizzard and Raymond Lewis. I was instructed to find out from the widows of the deceased individuals the amounts of money they owed on their homes. After I obtained this information, I went to Charleston and informed Blizzard and Lewis what each widow owed. These homes had been purchased from Island Creek Coal Company and were sold to them through the Aldredge Land Company. The balance due on Charlie Wells' home was $2,700; Robert Worley owed $2,200; Major Thompson owed $1,300 on the home he was living in and $2,700 on another nearby home that he had also purchased. This totaled $8,900 due on these homes. Raymond Lewis gave me $8,900 in cash with instructions to pay off the balance due on these homes. I was further instructed that I was to keep this transaction strictly confidential and in no way involve the organization.

Mr. Wells had two daughters approximately ten and twelve years of age. Mr. Worley had four children ranging from about six to twelve years of age, and his widow was in the family-way at the time of her husband's death. Mr. Thompson had three children ranging from about eight to fifteen years of age. The only source of income of the widows was Social Security benefits which the widows would receive until their last child reached the age of eighteen, and then Social Security would automatically stop.

In the meantime, Major Thompson contacted me and arranged for an appointment. When Thompson arrived at my office, he stated that he was disappointed in me because he had always had high regard and respect for me, and he had always considered me a close friend. I asked him what was wrong, and why he did not feel the same now. Thompson stated that he was very hard-up financially, because all he was receiving was his disability Social Security, and he could not make it on that. He was trying to get the union to give him some sort of pension to help him. He said that he and the committee at No. 1 Holden had taken the matter up with R. O. Lewis, secretary-treasurer of District 17 in Charleston, and Raymond Lewis had told Mr. Thompson that the reason he and the other two widows were not given any help by the union was because Mr. Blizzard and I objected to the union getting involved, and that was the reason Thompson had changed his attitude toward me. I then asked Thompson, "Do you really believe that, Major?" He said, "No, I don't believe it, but that is exactly what they told me." I responded, saying, "Major, do you remember when I came to your home and gave you the money to pay off your home? You said you owed $1,300 on it and you owed $2,700 on the other home you owned, making a total of $4,000, and when I gave the money to you, it was all in cash." Thompson responded, "Yes, you gave me that money to pay off those houses." I said, "Where do you think that money came from?" and he said, "I don't know," and I said, "Do you think that money came out of my own pocket?" I said, "Mr. Raymond Lewis gave me that money to give you to pay off those houses, and if he told you that Mr. Blizzard and I objected to helping you, he told you a lie. That is not true." I then advised Mr. Thompson to get his committee from No. 1 Holden, and if need be, I would go along with his committee to Washington to see if John L. Lewis would give him some help, and if Mr. Lewis attempted to tell him that the organization was not going to involve itself, Mr. Thompson should tell Mr. Lewis that I had given him $4,000 to pay off his home and that I received the money from R. O. Lewis.

Thompson convinced the committee to take him to Washington where they had an audience with John L. Lewis. Mr. Lewis told Thompson that he was sorry about his accident, but for him to go back home, and they would send him a check for $150 a month beginning the first of the month.

After Thompson received his first check for $150, he came back to my office to show me the check. He apologized again for believing what R. O. Lewis had told him about me and Mr. Blizzard being responsible for his not receiving any financial help from the international union.

This case is a prime example of how people can get involved in serious trouble and create a blemish on the record of the UMWA.

CHAPTER XIX

The Decade of the 1950s

The decade of the 1950s was a very successful period under the leadership of John L. Lewis. During this period, Mr. Lewis was able to raise the financial assets of the UMWA to its all-time high. By the UMWA assisting in the formation of the BCOA, and with the leadership of Harry Moses, formerly of U.S. Steel, as the head of BCOA, thus began an amicable relationship between the BCOA and the UMWA. This relationship brought a new era into the collective bargaining process in the coal industry, where successful negotiations were carried on that produced collective bargaining agreements in the coal industry from 1952 through 1964 without a single strike in the industry in this twelve-year period.

Harry Moses died on April 1, 1956. He was succeeded as president of BCOA on September 1, 1956 by Edward G. Fox who had been president of the Lehigh Coal and Navigation Company, an anthracite coal producer. Mr. Fox continued the method of man-to-man negotiations with Mr. Lewis through 1960 when Mr. Lewis retired from the presidency of the UMWA.

The early part of the sixties, brought forth some tragedies to the Lewis family with the death of Mr. Lewis's brother, A. D. "Denny" Lewis, who died of cancer. He had served as president of District 50 and was also John Lewis's most trusted lieutenant. Also, the death of John Lewis's daughter, Kathryn, former secretary-treasurer of District 50, who committed suicide and requested that her body be cremated, was quite a shock to Mr. Lewis because he was very close to his daughter. Several months later, his brother Howard committed suicide in Illinois. These tragedies occurred within less than a year of each other. In the early 1950s, two of Mr. Lewis's former close associates also died. Van A. Bittner, who succeeded Mr. Lewis as international statistician when Mr. Lewis became president of the UMW in 1916, former president of District 17 and vice-president of the United Steelworkers of America (USWA), died of cancer in Crafton, Pennsylvania. Phillip J. Murray, former UMWA vice-president and president of the USWA, who succeeded John L. Lewis as president of the CIO, died in a Pittsburgh hospital of heart disease.

It was reported in reliable circles in the UMW international office that several days prior to Phil Murray's death, Mr. Lewis learned that Phil Murray was in serious condition and that doctors held no hope for his recovery. Upon learning of Mr. Murray's condition, Mr. Lewis drove to Pittsburgh to the hospital where Mr. Murray was a patient and noted on the door of the room where Mr. Murray was located a sign that no visitors were allowed. Mr. Lewis saw the supervising nurse on the floor and requested that she go in and see if Mr. Murray would see him. When the nurse returned, she informed Mr. Lewis that Mr. Murray would not see him under any circumstances. After hearing the nurse's message, it was reported that Mr. Lewis sat down in a chair near the nurse's station and wept for several minutes and then left and returned to Washington. Upon returning to Washington the next day, he learned that Phil had died during the night.

Mr. Lewis attributed most of his success as president of the UMWA to practicing his philosophy of fostering nepotism within the UMWA. He had on the payroll: his brother Denny; his daughter, Kathryn; and, later, his youngest brother, Raymond, who later became vice-president of the union under President W. A. Boyle. He also had John Owens, secretary-treasurer of the UMWA; Owens' son, Willard, who was employed with the legal division of the UMWA; and Ronald Owens, who was secretary-treasurer of District 6. At one time, Owens and his sons received more than $100,000 per year in salaries. In District 19 in Jellico, Tennessee, he had William Turnblazer, president, and Turnblazer's son, William Turnblazer, Jr., who later succeeded his father as president of District 19. In District 30 in Lexington, Kentucky, there was Sam Caddy, president and his son, Sam Caddy, Jr., secretary-treasurer. Upon his father's death, Sam Caddy, Jr., succeeded his father as president. In District 15, in Denver, Colorado, was Frank Hefferly, president, who was later succeeded by his son, Frank Hefferly, Jr.

Mr. Lewis believed that by the practice of nepotism, the families would remain closer to him and carry out his mandates. He was criticized both by some members of his union and the media, but he paid no attention to what others thought about this matter.

After the death of his former friend, Phillip J. Murray, the second president of the CIO, a hard fight developed over who would succeed the late Phil Murray as president of the CIO. One faction was led by Walter Reuther of the United Automobile Workers Union (UAWU), and the other was led by Allan Haywood, director of organization for the CIO, who had the full support of David J. McDonald, the successor to Phillip J. Murray as president of the USWA. At the CIO convention in November 1950, Walter Reuther became the third president of the CIO by defeating Haywood in a close election. Haywood's support came from those unions who generally supported the views of the miners union and the late Phil Murray. Walter Reuther and his brother, Victor, were the sons of a machinist and one-time coal miner and were from Wheeling, West Virginia. The Reuther brothers spent some time working in an automobile factory in Russia and then came to Detroit and went to work in a plant of GMC. During the organization campaign of GM that was headed by Allan Haywood at that time, the Reuther brothers initiated a new type of strike in American industry by promoting the first sit-down strike in the plant of GM. The strike kept the plant closed until John L. Lewis came to Detroit and started negotiations with GM and Governor Murphy of Michigan. This led to the first contract in the automobile industry, and it was signed in Governor Murphy's office in Lansing, Michigan.

Nineteen fifty-two was quite a busy year in wage negotiations in the coal industry. This resulted in the amendment to the 1950 Wage Agreement which provided some major changes in the duties of the mine safety committee; the royalty provision for the 1950 Welfare and Retirement Fund was amended to increase the royalty payment from \$.30 to \$.40 per ton on coal; wages were increased \$3.50 per day; the vacation payment was increased to \$100.00 and there was insertion of a provision known as "Application of Contract to Coal Land." Other miscellaneous changes were also made. These amendments became effective October 1, 1952.

These amendments were negotiated between Mr. Lewis and Mr. Moses of the BCOA without any strikes in the industry.

During this period, as reported earlier, contracts were let out on one of the biggest projects ever undertaken by a labor union in this country—the building of ten miners memorial hospitals extending over the southern part of the Appalachian coalfields. One of the major hospitals was located in the south at Middlesboro, Kentucky, the second major hospital, was built in the central part of the Appalachian coalfields and located at Williamson, West Virginia, and the third hospital in the northern part of the Appalachian coalfields at Beckley, West Virginia. These three major hospitals were laid out so that the other seven hospitals would feed into them because they were able to cope with any type of surgery that a miner might have to have and conformed to the recommendations of Admiral Boone's survey of the coalfields.

These hospitals were completed by 1956, and Mr. Lewis was the speaker at the dedication in Beckley, West Virginia, to a throng of about 35,000 coal miners. It was one of Mr. Lewis's dreams come true, that of providing medical and hospitalization care to the miners of the Appalachian area. The cost of this enterprise was paid for out of the royalties paid by the coal operators into the UMW Welfare and Retirement Fund. Although transferred out of the UMWA management after thirty-five years of existence, these institutions continue to provide first-class medical and hospital services to the coal miners of the Appalachian region.

In seven short years, as a result of the successful negotiations of May 29, 1946 establishing the UMW Welfare and Retirement Fund, Mr. Lewis was able to mold the UMWA International Union into one of the most influential, prestigious enterprises in this country.

By 1953, the UMW Welfare and Retirement Fund was taking in approximately $200 million per year in coal royalties. These monies were on deposit in the National Bank of Washington. This bank was purchased in the middle of 1949 for $25 million for seventy-six percent of the outstanding common stock of the National Bank of Washington by the UMWA. The bank later became the third largest bank in Washington, D.C.

Mr. Lewis also invested in some of the promising real estate properties in Washington. The following properties were purchased by the International Union of the UMWA.

Location	*Cost*
907 15th Street, N. W.	$875,000.00
1427 Eye Street, N. W.	550,000.00
900 15th Street, N. W.	586,580.00
1523-5-7 Eye Street, N. W.	142,500.00
1435 K Street, N. W.	285,000.00
1437 K Street, N. W.	390,000.00
Total	$2,829,080.00

Mr. Lewis purchased these properties for two reasons. One, because the real estate was some of the best property in the city of Washington, and two, to accumulate enough fixed assets for the UMWA to make it the most powerful union in America. This added tremendous responsibility to Mr. Lewis. According to the audit for year ended June 30, 1960 by Wayne Kendrick and Company, Certified Public Accountants, Washington, D.C., the fund had surplus of $116, 271,505.29. In addition to his duties as president of the UMWA, Mr. Lewis was chairman of the UMW Welfare and Retirement Fund, and as a director of the National Bank of Washington, he exerted tremendous influence on the policy of the National Bank of Washington. He took pride in the financial status of the international union of the UMWA with its net assets at the beginning of 1960 of $110,320,531. With this vast accumulation of wealth owned and controlled by the UMWA, he was convinced his leadership had brought the union to its highest pinnacle and the union could not be destroyed by inside influences nor by influences from the outside. Mr. Lewis' influence was felt throughout the whole of American industry. He became intimately acquainted with some of the most outstanding leaders of the nation. His contributions to labor have never been equaled by his adversaries. When he organized the CIO in 1935, the numerical strength of all labor in this country was around 3,000,000. When he relinquished the presidency of the CIO in 1946, the numerical strength of the CIO alone was around 7,500,000.

Due to the rapid growth and expansion of organized workers in the basic industries of this nation under the banner of the CIO, Mr. Lewis had tremendous success organizing within the CIO. His accomplishments were unparalleled in the labor movement, and at the time of the AFL-CIO merger, there was a total of approximately 13,000,000 organized workers in America. This was the highest peak the labor movement ever reached. Today, the labor movement is at its lowest ebb, losing thousands of its members, and the labor movement in this half of the century has not developed the leadership to carry on the work that was started by such leaders as Samuel Gompers, John L. Lewis, William Green, Phillip J. Murray, Thomas Kennedy and George Meany.

Mr. Lewis was the most photographed person in America. His image was printed in cartoons in most of the newspapers of this nation. Neither before nor after Mr. Lewis's death has there been a person who was portrayed in the news media of this nation as much as John L. Lewis. It is hard to imagine a man who had less than an eighth grade education becoming America's greatest labor leader and also being America's most respected Shakespearean. During his lifetime, he took on anyone to defend the rights of the UMWA. He took issue with President Woodrow Wilson during World War I over not increasing the wages for the coal miners. He also took issue with President Harding and President Roosevelt over the government seizures of the coal mines in this country and broke relations with President Truman in 1946 over the government seizure of the coal industry. In these last two instances, the mine workers union was fined thousands of dollars, and in the latter instance, Mr. Lewis was personally fined $10,000 for contempt of court. Later, Mr. Lewis and President Truman became very close friends, and this relationship continued throughout their lifetimes. Mr. Lewis's greatest accomplishment in behalf of the UMWA came under government seizure of the coal industry in this nation.

John L. Lewis had the most disciplined organization in the history of the labor movement. This was due to his forty-six years of constant leadership, his unwavering influence with the membership of his union, and his day-to-day relationship with the leadership of the coal industry. He was an individual of great integrity—his word was his bond. He recognized the inherent rights of management in his dealings with the coal industry.

For many years, starting with the early days of 1933 when he organized the southern coalfields, he agreed in the national agreement to a provision known as the "Illegal Suspension of Work." This section of the agreement prohibited a miner from going out on an illegal work stoppage, and a miner would be fined two dollars per day for each day he participated in an illegal work stoppage. It also provided a section in the contract that prohibited a miner from suing in court to seek redress of any grievance he might have; and if the individual miner went to court, it was mandatory on the part of the district or the international union to defend the company in such action and provide counsel for the employer at the expense of the union. These provisions in the national agreement were of great benefit to the union and the members of the UMWA by forcing the membership of the union to utilize the grievance procedure as specified in the national agreement in seeking redress for their grievances. Mr. Lewis was very much opposed to illegal work stoppages and would not permit them during his leadership. The coal industry was grateful to him for recognizing this problem in the industry. Later, under government operation of the mines, these provisions were removed from the national agreement; however, Mr. Lewis was intelligent enough to recognize that since the union was new in the southern Appalachian coalfields, he still had to have provisions of this sort in the agreements so he could demonstrate to the coal industry that his union would honor the integrity of its agreements. His influence and power were recognized throughout the labor movement, business community, and the government of the United States. There is no doubt to those who knew him that he was the most dominant labor leader in the history of the American labor movement. Sometimes, he could also be a most ruthless individual, particularly to some of his closest associates that had crossed him in some manner or another.

CHAPTER XX

Labor's Hero

The late William Blizzard, president of District 17, Charleston, West Virginia, was for a long time one of Mr. Lewis's closest and most trusted lieutenants. This relationship covered a span of more than forty years. Mr. Blizzard, at an early age of twenty, became an active organizer for the UMWA, particularly in the Cabin Creek district in Kanawha County, West Virginia under President Frank Keeney of District 17 and Secretary-Treasurer Fred Mooney.

During the fall of 1921, Red Jacket Coal Company, the largest coal company in Mingo County, decided it would operate nonunion, and it notified the UMWA employees who lived in the company houses to vacate their property. This, the miners refused to do. The company engaged the Baldwin-Felts Detective Agency in Bluefield, West Virginia to come to Matewan, West Virginia to evict the miners from their homes. Sid Hatfield, the chief of police of Matewan, sent word to Tom Felts, head of the detective agency, that if they came into Matewan on the train to evict the miners, he would not permit them to get off the train. Tom Felts ignored the chief of police's instructions and sent several of his detectives to Matewan on a No. 3 train that arrived at the Matewan train station in early afternoon. When Felts' detectives got off the train, they were met by Sid Hatfield and Mayor Testerman of Matewan. They were advised not to get off the train. After ignoring this advice and leaving the train at the station, shooting began. When the shooting stopped, several of Tom Felts' detectives were dead along with one of his brothers, Mayor Testerman of Matewan, and a miner. Sid Hatfield and Ed Chambers, a miner, were indicted for this crime in the Mingo County Circuit Court, Williamson, West Virginia. Due to the wide notoriety of this incident, Hatfield and Chambers agreed to a change of venue to McDowell County at Welch, West Virginia. The sheriff of Mc-

Dowell County was Greenway Hatfield, and he assured both Sid Hatfield and Ed Chambers that they would be fully protected. On the morning they were to be tried, Sid Hatfield and his wife, Ed Chambers and his wife, and the UMWA Chief Counsel, Harold Houston of Charleston, West Virginia, were going up the steps of the McDowell County Courthouse about 8:30 A.M. when Sid Hatfield and Ed Chambers were both killed with 30-30 high-powered rifle shots.

The news of the deaths of these two men spread like wildfire through the unionized coalfields of Kanawha County, and this infuriated the miners in Kanawha County. About ten thousand miners in Kanawha County decided to march into Mingo County to Williamson to assist the miners of Mingo County. This march was organized by Frank Keeney and Fred Mooney, and they placed William Blizzard in charge of the march. Blizzard and the miners march got as far as Blair Mountain in Logan County when they were encountered by the deputies of Don Chafin, sheriff of Logan County. There were several shots fired resulting in the death of one of Sheriff Chafin's deputies, John T. Gore; together with John Muncie, tipple foreman of the Monitor Coal and Coke Company; Wilkinson, West Virginia; and J. W. Swanner, mine foreman of the D. H. Pritchard Coal Company, Ethel, West Virginia. As a result of the murders of these men, Blizzard was lodged in the Logan County jail with the charge of murder against him. Keeney and Mooney had hidden out in Columbus, Ohio and did not participate in this march. Blizzard remained in the Logan County jail until he was indicted for murder and treason, but his attorney was granted a change of venue to Fayetteville. At the Fayetteville trial, he was charged with murder and treason. The jury was unable to agree on a verdict, and the case was transferred to Lewisburg, Greenbrier County. Again, the jury was unable to reach a verdict. Later, the case was transferred to Charles Town in Jefferson County where he was tried for treason. This was in the same courthouse where abolitionist John Brown had been tried and subsequently executed.

After the Fayetteville and Lewisburg trials which resulted in hung juries, T. C. Townsend was employed by the UMW International Union to represent Mr. Blizzard. When Mr. Townsend learned that the trial was being transferred and scheduled in Charles Town, Mr. Townsend recognized that the scheduling of this case was influenced by the coal operators who were convinced that Blizzard would be convicted there, since Jefferson County was totally a farming area. In those days, the farmers had no sympathy for the union nor anyone connected with them. Mr. Townsend employed about twenty-five salesmen and purchased about twenty-five hundred Bibles and distributed them among the salesmen. These salesmen went into Jefferson County three months before Blizzard's trial and visited all areas of Jefferson County, selling Bibles to the farmers. After they delivered their religious pitch to the farmers concerning the purchase of the Bible, they would tell the farmers how mean the coal operators were in southern West Virginia, and that the operators would not permit people selling Bibles on their properties, thus poisoning the minds of these farmers against the coal operators. This scheme worked quite successfully for Mr. Townsend because when the jury was selected, most of the jurors had prior knowledge of the coal operators in southern West Virginia's inhumane treatment whereby they were depriving the workers of their constitutional rights. Thus, the jurors minds were already made up, and when the verdict was handed down, they ruled in favor of Blizzard and said he was "not guilty" of the crime he was charged with.

Blizzard told me this story several times. As a result of this scheme by Mr. Townsend and the fact that he had gotten Blizzard acquitted, after the trial he was employed as chief counsel for the UMW in West Virginia and continued as its chief counsel until his death in the early 1950s.

The trial lasted for several weeks, and Mr. Lewis attended the trial. Blizzard was acquitted of the treason charges by the jury. There was wide coverage of this trial by the media because Blizzard was the only man tried for treason in this country other than John Brown.

"But, after his acquittal and he returned to Charleston, he was met with a parade that marched up Capitol Street. He was tried, again, this time for murder instead of treason. The case in Greenbrier County ended in a hung jury, and Blizzard once again was the miners' hero.

"In comparing his treason trial once to Brown's, Blizzard said: 'We both wanted to free the slaves.'

"Blizzard served as vice president of District 17 for many years under Bittner and Percy Tetlow, and was named president in November of 1945.

"From 1922 until his retirement, he was the UMW's legislative representative of the State Legislature where he held those duties. He lived to see most of the measures for which he and others had fought enacted into law."[1]

After Blizzard was freed of these charges, he became a very close associate and trusted lieutenant of Mr. Lewis, until the early part of January 1955, when John Lewis's brother, R. O. Lewis, who at that time was secretary-treasurer of District 17, changed the practice of delivering the mail at the district office. It was customary practice that the secretary-treasurer received all of the mail in the office, and it was sorted out and delivered to all of the departments and offices of the UMW, and for some reason or another, all of Mr. Blizzard's mail was being opened before being delivered to his office, including his personal mail. Blizzard went down to Raymond Lewis's office and advised him that he did not object to the opening of the mail of the UMW, but he did object to his office staff opening his personal mail. He requested that this practice be discontinued. A few days later while Mr. Blizzard was attending a session of the Legislature at the State Capitol, Mrs. Waldo, Raymond Lewis's secretary, went to Mr. Blizzard's office with his mail, and she had a letter addressed to Raymond Lewis from Mr. Blizzard. The substance of the letter was a receipt that Mr. Blizzard had received his mail and an acknowledgement that his personal mail had not been opened. Since Mr. Blizzard was not in his office that morning, Mrs. Waldo requested that Mrs. Burke, Mr. Blizzard's secretary, sign and accept Mr. Blizzard's mail. She refused and said she would have to bring the mail back when Mr. Blizzard came in the next morning. So, Mrs. Waldo took the mail back to Raymond Lewis's office. The next morning, Blizzard had a meeting in his office with several of his subordinate officers, and Raymond Lewis came up to his office and handed him the letter and requested that Bill sign it. After Bill read the letter, he told Raymond he did not have to sign for his mail; that his mail belonged to him, and he would not sign for it. This developed into an argument. One of Raymond Lewis's favorite expressions was "you are an s.o.b." When he called Blizzard this, Blizzard hit him with his fist between the eyes; knocked him to the floor; and blackened both of his eyes. The next day, the *Charleston Gazette* carried the story that a certain labor leader was sporting two black eyes. As a result of this altercation between Raymond Lewis and William Blizzard, John L. Lewis wired Blizzard and requested that he meet with him in Washington on February 11, 1955.

1. The *Charleston Gazette*, August 1, 1958.

When Mr. Blizzard met with Mr. Lewis in his office, Mr. Lewis pointed out to Blizzard that he had the most colorful labor-leader career in the United States; that he had worked hard during his lifetime in the best interests of labor and the UMW in particular; and, since he had this outstanding record, he advised Blizzard that he had arranged with John Owens, the international secretary-treasurer, to put Blizzard on a pension starting April 1, 1955, paying him $400 per month. Blizzard decided that if he did not take the pension, Mr. Lewis would transfer him to some area of the country where there was no organization, and he would have to leave his employment to get back home. So, Blizzard decided to take the $400 and retire. After he agreed to accept Mr. Lewis' proposal, Mr. Lewis buzzed his secretary, Mrs. Covington, and she came into the office with her stenographic book. Mr. Lewis had Mr. Blizzard dictate his retirement letter to Mrs. Covington. Before Mr. Blizzard could get back to Charleston, Mr. Lewis had already mailed this letter of Blizzard's retirement throughout all local unions in District 17. This was a bitter pill for Mr. Blizzard to swallow.

During the conference discussing his retirement, Mr. Blizzard began telling Mr. Lewis what had happened between him and Raymond. Mr. Lewis responded that he didn't give a damn what had happened, so Blizzard said no more. Blizzard returned to Charleston to his office and was quite upset with the manner in which Mr. Lewis had treated him. Blizzard received his first pension check on the first of April in the amount of $400. His second check, which would have been May 1, was sent to him with a letter from Mr. Lewis advising Blizzard that, due to economic conditions which existed at that time, his pension would be reduced to $300 per month. Blizzard continued to draw the $300 per month pension until about ten days prior to his death. Blizzard who died of cancer on July 31, 1958, had been treated for cancer at the National Institute of Health, and when Mr. Lewis found out that Blizzard's death was near, he sent a letter to Mr. Blizzard advising they were reinstating the $100, bringing his pension back to $400 per month.

A few days prior to his death, Mr. Blizzard's wife called me. I was in Tazewell, Virginia, at the time. She told me that her husband had just a few days to live, and if I wanted to see him before he died, I had better get down to see him at once. I, along with a close friend of Blizzard's left the next day to see him. He showed me the letter he had received restoring the $100 to his pension. Blizzard was very much displeased, and he told me that if he had known he was going to receive the kind of treatment he received working for the UMW, that he would not have made the sacrifices he did during his lifetime for the union. He went to his grave a very saddened man because of the treatment he received from Mr. Lewis.

I worked for Mr. Lewis for eighteen years and knew him well. I can recall the same treatment Mr. Lewis gave Phil Murray when he removed him as vice-president of the UMW. At times, Mr. Lewis was a very ruthless man, and he did not permit anyone to stand in his way.

CHAPTER XXI

John L. Lewis's Retirement

In the latter part of 1958, Mr. Lewis gave his full attention to working out a new wage agreement with the BCOA and its new president, Edward Fox, who had succeeded the late Harry Moses. The new agreement became effective December 1, 1958, amending the 1950 National Wage Agreement. The new agreement provided for a $2.00 per day wage boost; a $20.00 increase in vacation pay to $200.00 for 14 days; and a "protective wage clause." The first part of the wage boost—$1.20 became effective January 1, 1959. On April 1, 1959 another $.80 per day was added. The increase brought the union coal miners basic wage to $24.25 a day in the Appalachian field.

The agreement was signed by BCOA on December 3, 1958, and by the Southern Coal Producers Association (SCPA) on December 4, 1958. It was an "open end" agreement running until November 30, 1959, after which it could be terminated by either party by giving sixty days notice. However, this agreement continued through the early part of 1964.

This was the last official agreement negotiated between Mr. Lewis and Edward Fox, president of the BCOA. At this point in time, Mr. Lewis was (in his own mind) feeling he had accomplished all that was to be accomplished under his leadership on behalf of the mine workers in the United States and Canada. His leadership as president had spanned a period of approximately forty-five years, and he was convinced that the UMWA had reached its pinnacle and that his leadership had molded the organization into being the greatest labor organization the world had ever known. By then, the UMW's power and influence, both financially and politically, on the North American continent was unequalled by any other labor organization.

The union's total assets had reached $110,320,531 at the beginning of 1960, and some of Mr. Lewis's most outstanding accomplishments were the complete organization of the coal industry under the banner of the UMWA; a wage agreement providing for industry-wide bargaining; the establishment of the union shop; elimination of the forty-cent wage differential between the north and the south; portal-to-portal pay; the establishment of the UMWA Welfare and Retirement Fund; and the construction of ten hospitals in the Appalachian area to improve the health and welfare of the miners and their families. His greatest disappointment was that he was unable to contribute his talents to unifying the labor movement. With all of these achievements, Mr. Lewis was convinced that his union had reached its peak and that it was time for him to retire and relinquish the reins of leadership to his closest friend and associate, Vice-President Thomas A. Kennedy.

At a special meeting of the International Executive Board of the UMWA on January 14, 1960, Mr. Lewis submitted his retirement to the board and recommended that the board designate Vice-President Thomas A. Kennedy to succeed him as president and that W. A. Boyle, Mr. Lewis's administrative assistant, be promoted to vice-president. Mr. Owens was to continue as secretary-treasurer.

During the tenure of Mr. Lewis as president of the UMW and chairman of the UMW Welfare and Retirement Fund, he made available special considerations to the friends of the UMW organization to assist them in securing loans from the National Bank of Washington to benefit the UMW organization such as in the case of J. P. Routh, chairman of the board and chief executive officer of the Pittston Company. Mr. Routh received monies from the National Bank of Washington through the influence of Mr. Lewis to purchase the properties and holdings of Colonel J. G. Bradley, who owned the largest nonunion coal operation in Clay County, West Virginia. This included the Elk River Coal and Lumber Company mines, which was 91,000 acres on Buffalo Creek and included the entire watershed and extended into Nicholas and Braxton counties. This acquisition of Colonel Bradley's properties was completed in 1957; these properties were then unionized.

Shortly after the sale of these properties, Colonel Bradley, at the age of seventy-six, left West Virginia and returned to Needham, Massachusetts ending a lifelong career of operating nonunion mines in West Virginia.

Colonel Bradley died in March of 1971 at the age of eighty-nine.

This acquisition of Colonel Bradley's 91,000 acres of coal properties in Clay County, West Virginia, along with Cyrus Eaton's acquisition of the West Kentucky operations—and their subsequent coming under contract with the UMWA—

brought to an end the last two giants of nonunion operations in the bituminous coal industry.

Under his forty-year leadership, the accomplishments of Mr. Lewis for the mine workers of the United States and Canada are monumental and well-known throughout the mining industry, as well as his outstanding contribution of leadership in the labor movement, particularly, for those unions in the automobile industry, steel industry and the rubber workers, and many other unions who came under the banner of the CIO. They owe their existence to Mr. Lewis personally. No doubt, to those who knew him, he was by far the greatest labor leader the world has ever known.

In spite of all of these accomplishments in behalf of the mine workers of this nation and labor in general, he has, however, never been memorialized by the mine workers and labor. It has been left to the residents of the town of Lucas, Iowa, where Mr. Lewis was born and raised, to come forth to establish a museum in his memory. The following article is quoted from the Huntington *Herald-Dispatch* and contains more information:

> Millions of people who never saw a coal mine or a miner know who John L. Lewis was.
>
> Oddly enough, here in Lucas, the town of his birth, no one ever made a big deal of Lewis, who headed the United Mine Workers union for 40 years and was instrumental in the founding of the Congress of Industrial Organizations.
>
> That's about to change.
>
> On Saturday, ground will be broken for the John L. Lewis Mining and Labor Museum.
>
> "It's going to give the people in town a lot of pride," said Jean Hager, a member of the John L. Lewis Commission, the group that has spearheaded the project.
>
> Ms. Hager and others feel the museum will honor the memory of the town's most famous son, who became a giant in the American labor movement before his death in 1969 at age 89.
>
> They also think the museum will boost the town, which has been going downhill since the coal mines were closed nearly 70 years ago. The population is 300. It was about 2,000 when Lewis was born there in 1880.
>
> Kathryn Dixon, another commission member, said at least two other attempts to build a memorial failed, chiefly because no one bankrolled the project.
>
> This time the Lucas folks got the boost they needed when the Iowa Legislature approved a grant of $40,000. The United Mine Workers Union kicked in $11,000. Other union groups from the AFL-CIO have contributed, and local folks, through

pledges and from the sales at a farmers' market and craft shop, contributed about $16,000.

"We're still about $14,000 short," said Ms. Hager. But the project will proceed, she said.

Plans call for a 70-by-75 foot building to house such things as tape recordings of Lewis' speeches and a documentary film.

It will also have such memorabilia as an early mine union ledger showing Lewis' payment of dues and various mining tools.

The type of power Lewis wielded as a union leader may never be seen again. In the 1940s and 1950s, Lewis—with a word to his union—virtually could shut down the American industrial economy. When the coal miners walked out of the pits, it wasn't long before the steel industry, which depended on coal, was jeopardized.

Lewis filed court orders and quarreled with presidents. He was hated by some but loved by the miners who saw their pay rise under his leadership from $2 to $24.25 a day, plus generous pension and medical benefits.

"A memorial to this man is long overdue," said Ms. Dixon. "A lot of people didn't like him. But you can't do what he did and not make enemies."

Ms. Hager said most townspeople hope the museum sparks other development—a restaurant, for example. Just getting the project off the ground has spurred others to knock down abandoned houses and buildings, she said.[1]

In my view, it is a disgrace for the UMW organization to contribute only a pitiful $11,000 toward the establishment of this museum and the small contribution by the AFL-CIO. If they really appreciated what Mr. Lewis had done in regard to improving the fate of the mine workers and labor unions during his reign as their leader, they would provide all of the money without this committee having to solicit it to establish such a monument for his leadership. This doesn't speak well of their appreciation.

The International Executive Board unanimously approved the recommendations of Mr. Lewis for the new slate of officers, and his retirement became effective January 14, 1960. President Kennedy recommended that President Lewis be given the title "President Emeritus" and also be given office space in the UMW building during his lifetime. This was unanimously approved. However, Mr. Lewis did not resign his position as chairman and trustee of the UMW Welfare and Retirement Fund, but continued in this capacity until his death.

1. The *Herald-Dispatch*, Huntington, West Virginia, April 1, 1989.

President Lewis spent the remainder of his lifetime going to his office on the fifth floor of the UMW building almost daily. His many friends and associates in the coal industry, government, including congressmen and senators, whenever they had some spare time, would drop by Mr. Lewis's office and discuss some of the current events of the day and some of the events of his past. He really enjoyed his retirement, with the exception of some of the events that happened after Mr. Boyle assumed the presidency of the union in 1963.

Mr. Lewis was admitted to Doctor's Hospital in Washington, D.C., and on the evening of June 11, 1969, the coal miners of the nation received the sad news that he had died. He was eighty-nine years old. Messages of condolence and sympathy were received by his family from all walks of life; from President Nixon, President Nixon's Cabinet officers, prominent personalities, both in and out of government, and labor organizations, including a message from George Meany, President of the AFL-CIO.

Two of the most outstanding messages of sympathy were President Nixon's and George Meany's. Following is the text of these two messages:

President Nixon

> The death of John L. Lewis takes away from the American scene one of the giants of our time, a man who helped to shape modern America. His talents were great, his energies prodigious. Few men in contemporary American life have been as controversial; few men possessed the proud heart and the passionate mind to engage in such large-scale controversies. His impact on the coal mining industry, on organized labor, and on the entire range of the American economy cannot be exaggerated.
>
> Mrs. Nixon joins me in extending deepest sympathy to his family.[2]

George Meany

> Death has removed a strong pillar from the House of Labor. John L. Lewis was truly a giant in spirit and in accomplishment. He offered strength to those who were weak. His courage inspired those who followed him to victory in the struggle for social justice. His entire life was a crusade for the rights of all workers. The AFL-CIO joins with all Americans in mourning the passing of John L. Lewis.[3]

2. *United Mine Workers Journal,* June 15, 1969, p. 13.
3. Ibid., p. 13.

At Mr. Lewis's request, his body was cremated. President Boyle called for a memorial service to be held throughout local union halls and churches in the United States and Canada in memory of Mr. Lewis. The full text of Boyle's message upon learning of Mr. Lewis's death is as follows:

The thundering voice of America's greatest labor leader has been stilled. The passing of John L. Lewis is a great loss for the United Mine Workers of America and a personal tragedy for me. I have lost a friend, a counselor and mentor. John L. Lewis brought me from Montana, guided me and helped me to carry the heavy burdens I now must carry. His passing leaves a void in my life that can never be filled.

To the working men and women of the world, he was the symbol of dignity, strength, unity and labor's struggle for betterment. To America's coal miners, he was the fighter who led them from serfdom to their rightful place as first-class members of the American society.

John L. Lewis was a man of deep compassion, possessed of tremendous capabilities, extraordinary vision, sheer genius of mind and, as characterizes his brother coal miners, had indomitable courage. His entire life was dedicated to improving the lot of his fellow man. His name is a legend, as is the inspiration he has given to all of us. All America and the entire free world mourn his death. But it is felt most deeply among the members of the United Mine Workers of America and its leadership.

I am asking that on Monday, June 16, 1969, all coal miners gather in their respective churches and union halls for services in memory of John L. Lewis. Beginning at 12:01 A.M., June 13, 1969, until after the funeral there will be a period of mourning during which time all coal mining will cease in the United States and Canada as we honor our fallen leader. It is altogether fitting that the coal mines be silent while the men who work in them come together to do honor to John Llewellyn Lewis.

(On June 13, 1969 Boyle advised UMWA members that the funeral was private with a memorial service to be held later. The mourning period in the coal industry was to end at 12:01 A.M. Tuesday, June 17.)

As an even more fitting memorial to him, I demand that the United States Congress immediately pass strong coal mine health and safety legislation. A John L. Lewis health and safety bill would be a fitting climax and memorial to the career of this outstanding champion of coal mine safety.

We hope that Americans of all stations and especially the American labor movement will now rally and support the push of John L. Lewis' union for health and safety legislation.

John L. Lewis best described himself at a convention of coal miners when he said: 'I have pleaded your case from the pulpit and public platform . . . not in the quavering tones of a feeble mendicant asking alms, but in the thundering voice of the captain of a mighty host, demanding the rights to which men are entitled.'

That was John L. Lewis.[4]

Thus ended the life of John L. Lewis, a great giant in the annals of the labor movement and the United States of America.

4. *United Mine Workers Journal*, June 15, 1969, p. 3.

CHAPTER XXII

The Thomas Kennedy Years

The newly elected president of the UMWA, Thomas Kennedy, was Mr. Lewis's closest associate and had been for more than forty years. Mr. Kennedy was from the anthracite fields, and his home was Hazelton, Pennsylvania. In 1924, when Samuel Gompers, president of the AFL died, John L. Lewis, who was then president of the UMW and also a vice-president of the AFL, proposed to the Executive Council of the AFL that William Green of Coshocton, Ohio, former president of District 6, UMWA, Columbus, Ohio, then serving as international secretary-treasurer of the UMWA, be elected to succeed the late Sam Gompers as president of the AFL. The Executive Council unanimously elected William Green as the new president of the AFL, becoming effective immediately. Thomas Kennedy succeeded William Green as secretary-treasurer of the UMW in 1924. Mr. Kennedy later succeeded John J. O'Leary as vice-president of the UMWA in 1947.

For more than forty years, Thomas Kennedy was John Lewis's closest friend and associate. Mr. Lewis always referred to Kennedy as "Governor" because in the Roosevelt landslide of 1932, the state of Pennsylvania elected its first Democratic Governor, George Earle, and Thomas A. Kennedy was elected Lieutenant Governor, the first time a labor leader had been elected to that high office. Kennedy and Earle were both reelected in 1936 for a second term. Kennedy did not give up his office in the union and continued to serve as secretary-treasurer. After John L. Lewis, Mr. Kennedy was the most qualified person to ever ascend to the presidency of the UMWA, because of his more than 40-year association with John L. Lewis. Mr. Kennedy's term of office as president of the UMWA extended from January 14, 1960 until his death on

January 19, 1963, at his home in Hazelton, Pennsylvania. His death was attributed to cancer.

Since Kennedy's tenure as president of the UMW was of short duration, he did not have an opportunity to negotiate a contract in the bituminous coal industry during his term as president of the UMW, but he had been in all of the wage negotiations of Mr. Lewis from 1924 through 1960. Kennedy, during his lifetime, negotiated all of the contracts for the anthracite industry, of which he was more familiar. The anthracite operators had the utmost respect for Kennedy's character and integrity in his dealings with them, and the operators affectionately called him "The Little Giant of the Anthracite."

Mr. Lewis did not attend the funeral, even though Tom Kennedy, I would say, was his most trusted and closest associate. Their deep friendship existed for more than forty years. For some unknown reason, I never knew Mr. Lewis to attend any of his closest associates funerals in all of his many years as president of the UMW.

CHAPTER XXIII

The Boyle Years

After Kennedy died, W. A. "Tony" Boyle, the international vice-president, assumed the presidency of the union.

Mr. Boyle had been a coal miner in the Billings, Montana area when he was elected president of District 27 with headquarters in Billings, Montana. He served for a short period in that capacity and was later assigned by President Lewis as the regional director of the CIO for the states of Oregon and Washington. He had been active in contending with the rival Mine, Mill and Smelter Workers Union in efforts to organize western miners and had been severely beaten in one of his confrontations with that union. When the split came between Phil Murray and John L. Lewis in 1940, Boyle decided to leave the CIO and come back to the UMW. Mr. Lewis assigned him to work out of his office. Boyle served as an assistant to Mr. Lewis until January 1960, at which time he was designated vice-president.

However, before Vice-President Boyle could be installed as president of the UMW, Mr. Lewis made it known to Mr. Boyle that he desired that his brother, Raymond O. Lewis, who at that time was secretary-treasurer of District 17, UMWA, Charleston, West Virginia, be named vice-president of the International union. The International Executive Board approved the new vice-presidential selection unanimously.

Raymond Lewis began his employment with District 17 in Charleston, West Virginia in 1946 as the district auditor. Prior to that, beginning in 1935, he had been employed as custodian of equipment of the Work Progress Administration (WPA) for the state of West Virginia, with headquarters in Clarksburg, West Virginia.

During the senatorial campaign in 1934 in West Virginia, John L. Lewis and Van A. Bittner, then president of District 17, were very successful in nominating and electing Rush D. Holt,

Democrat, from Weston, West Virginia, to the U.S. Senate. Rush Holt was the youngest senator ever elected to the U.S. Senate. He was only twenty-nine years of age at the time of his election. He had to wait until his thirtieth birthday to qualify to take the oath of office. In the early part of 1935, when he became thirty years of age, he qualified and was sworn into office as the junior senator from West Virginia. During this period, there was no Hatch Act, a law to prohibit government employees from participating in political campaigns, and the senior senator from West Virginia, M. M. Neely, controlled all of the federal patronage in West Virginia.

Matt Holt, Senator Holt's brother, held a position as custodian of equipment for the WPA, which was one of President Roosevelt's emergency programs to put people back to work, with headquarters in Clarksburg, West Virginia. All state equipment for this program was under his supervision. Raymond Lewis, prior to coming to West Virginia in 1935, was unemployed and more or less traveled throughout the United States from one place to another. He finally decided he would like to come to West Virginia to work. He discussed this matter with John Lewis, his older brother. John Lewis prevailed upon Senator Neely to give Raymond Matt Holt's job as custodian of the WPA equipment for West Virginia. So, Senator Neely had the state administrator of the WPA dismiss Matt Holt and put Raymond Lewis in his place. After Matt was dismissed, Senator Holt broke his allegiance to John Lewis and all during his senate career, he opposed John Lewis and opposed all labor legislation that came through the senate while he was a senator.

Raymond Lewis continued to work for the WPA as custodian of equipment through 1940, when the WPA was abolished and the National Defense Program began. The National Defense Program was created to prepare America for World War II which was already spreading in Europe. In the meantime, Senator Neely was elected governor of West Virginia and was sworn into office on January 1, 1941. He selected, as his State Road Commissioner, Ernest L. Bailey. The new road commissioner was a mining engineer and had formerly operated coal mines in the Mingo County, West Virginia area. Governor Neely requested Road Commissioner Bailey to place Raymond O. Lewis in charge of the state road equipment and for Raymond Lewis to be solely responsible to the road commissioner for his work.

Sometime later, in the early part of the Neely administration, Commissioner Bailey and Raymond Lewis began to be at odds with each other. This situation continued over a period of time, and finally, the road commissioner went to the governor's office and discussed the situation with Governor Neely. The governor would not advise Bailey as to what action he should take in regards to Raymond Lewis. The commissioner knew full well that if he took any action in dismissing Raymond Lewis, the administration would be at odds with John L. Lewis and the UMWA. This situation continued during the life of the administration of Governor Neely.

On January 1, 1945, Governor-elect Clarence Meadows was inaugurated as the new governor of West Virginia. He retained Commissioner Bailey as the State Road Commissioner in his administration. Shortly after Governor Meadows was settled into office, Commissioner Bailey went to the governor's office to discuss the Raymond Lewis situation. He pointed out to Governor Meadows that he had an employee in his office who would not keep his mouth shut, and he had to do something about the situation. Governor Meadows responded by saying that if he had someone working in his office who would not keep his mouth shut, he should fire him. That is what Commissioner Bailey wanted the governor to say. The commissioner went immediately down to his office and fired Raymond Lewis. For approximately eight or nine months, Raymond Lewis was unemployed and the word went out from John L. Lewis's office to the leadership of the mine workers in West Virginia that Clarence Meadows was anti-United Mine Workers. The union representatives were instructed to oppose Meadows' programs. Sometime after Raymond Lewis was dismissed from his employment with the road commission, John Lewis sent his brother, Denny, down to Charleston to see Bill Blizzard, president of District 17, and he told Blizzard that John Lewis had sent him down to see if he could find a position in the district organization to employ Raymond. Blizzard put Raymond to work as an auditor for District 17, auditing local union records. There never had been a district auditor in District 17 before nor after Raymond Lewis served in this capacity.

After Blizzard agreed to employ Raymond, Blizzard recognized the fact that Raymond was not a member of the UMW and the constitution required anyone employed by the UMW to be a member in good standing with the UMWA. In order for Blizzard to protect himself from any criticism for employing Raymond Lewis, he gave Jim Farley, who was in charge of the Kanawha District of the UMWA, ten dollars, with instructions for Farley to go to Harewood local union on the Kanawha River to pay Raymond O. Lewis's initiation fee and six months dues. The president of the local union, Herman Adkins, agreed to accept the initiation fee and six months dues. This automatically placed Raymond Lewis in good standing with the local union; however, he still did not qualify under the constitution of the UMWA, because the constitution required that a person who was employed by the UMW had to have at least five years experience working in the coal mines in this country, and Raymond Lewis had no record of ever having worked in the coal mines. Later, Denny Lewis, John Lewis's administrative assistant and brother, recognized the fact as to how Raymond had become a member of the UMWA, and he was not too sure that Herman Adkins would keep this matter confidential. However, in order for Denny Lewis to be assured that Adkins would not say anything about this matter, he gave Adkins a job as a representative of the United Construction Workers Union, an affiliation of District 50, UMWA. Adkins, in due course, became a very trusted lieutenant of Denny Lewis.

Upon the death of Thomas Kennedy in January 1963, the total assets of the union were $92,078,268. W. A. "Tony" Boyle, vice-president of the union, became the successor to Thomas Kennedy, and Raymond O. Lewis was designated vice-president of the international union, succeeding W. A. Boyle.

After his appointment as vice-president of the International Unión, Raymond Lewis began to disagree with the policies of W. A. "Tony" Boyle. Finally, Boyle proposed to Raymond Lewis that if he would retire as vice-president of the international union, Boyle would see that he would continue to receive his vice-presidential salary. Raymond Lewis agreed to accept Boyle's offer and retired and moved to Florida.

Raymond Lewis continued to live in Fort Myers, Florida from 1964 until the middle of October of 1987 at which time he died at the age of eighty-seven as a result of a stroke he had suffered. Raymond Lewis was the last surviving member of his immediate family.

After Raymond Lewis's retirement, George J. Titler, president of District 29, UMWA, Beckley, West Virginia, was named by Boyle to succeed Raymond Lewis as vice-president of the International Union. Titler had been one of John Lewis's trusted and loyal lieutenants.

John Lewis never accepted Boyle's reasoning for ousting his brother. After this move by Boyle, the relationship between

John Lewis and Tony Boyle gradually grew astray. I recall, on one of my visits to John Lewis's office on the fifth floor of the UMW building, that Mr. Lewis expressed his resentment of Boyle, but since he was getting up in years, he did not think it would be wise for him to openly express his resentment of Boyle. This confirmed what I had heard—that Mr. Lewis had confided in some of his closest associates the fact that he was unhappy with the leadership of Mr. Boyle. However, Mr. Lewis did not live long enough to witness Mr. Boyle's further deteriorating leadership and mismanagement of the affairs of the union, or his disgrace in being ousted as president of the UMWA.

Shortly after Raymond Lewis retired as vice-president, Mike Widman, the director of the department of research for the UMWA, became disenchanted with Boyle's leadership, and he retired and returned to his home in Springfield, Illinois.

Mr. Widman was a very capable and intelligent individual. In 1936, President John L. Lewis assigned Mr. Widman the task of organizing Ford Motor Company. Ford Motor Company was the largest employer to be organized in the automobile industry, and Widman was given full credit for a successful organization campaign. Widman was originally a coal miner from southern Illinois, and due to his ability as an organizer, he was one of Mr. Lewis's most trusted lieutenants. After the death of Mr. Lewis's wife, Mr. Widman purchased the home of John L. Lewis in Springfield, Illinois, where he returned upon his retirement from the union.

Many of Mike's closest associates in the UMW were open in their expressions that he should have been the one to have succeeded the late Thomas Kennedy as president of the UMWA.

In the meantime, Boyle continued to have problems with his leadership of the union. Welly Hopkins, who was general counsel for the UMWA, also had difficulty in getting along with Mr. Boyle, and Boyle terminated his services as general counsel for the UMW. A short time later, when Mr. Lewis found out that Mr. Hopkins had been relieved, Mr. Lewis employed Hopkins as general counsel for the UMW Welfare and Retirement Fund.

Mr. Lewis continued to serve as chairman of the UMW Welfare and Retirement Fund after his retirement as president of the UMW. He served as chairman of the fund until his death on June 11, 1969. At the time of Mr. Lewis's death, the Welfare and Retirement Fund had paid more than $2.6 billion in benefits to nearly five million beneficiaries.

In 1964, Boyle and the president of the BCOA, Ed Fox, negotiated an amendment to the 1950 Wage Agreement running to March 31, 1966. At the expiration of the agreement in 1966, the 1950 Wage Agreement was amended again to run until April 1, 1968.

On September 1, 1967, Edward G. Fox retired as president of BCOA. During Mr. Fox's tenure, he made many friends throughout the coal industry, many of whom he continued to keep in touch with after his retirement on September 1, 1967. After his retirement, Mr. Fox returned to his home in Pottsville, Pennsylvania, and later died on October 12, 1987.

Mr. Fox was succeeded by George L. Judy, a vice-president of Consolidation Coal Company's Christopher Coal Division. Prior to his position as vice-president of Consolidation Coal Company, Mr. Judy had been president of the Pittston Company's Clinchfield Coal Division in Dante, Virginia.

Upon the retirement of Ed Fox and the succession by George Judy, the International Union of the UMW returned to the old policy of negotiating wage agreements by its international officers with the Executive Committee of BCOA. The BCOA committee consisted of John Corcoran, President and CEO, Consolidation Coal Company, Pittsburgh, Pennsylvania; R. Heath Larry, Vice-Chairman of the Board of U.S. Steel, Pittsburgh, Pennsylvania; Ed Phelps, president, Peabody Coal Company, St. Louis, Missouri; and Herbert E. Jones, Jr., President, Amherst Coal Company, Lundale, West Virginia.

This ended an era of two-man negotiations which lasted from 1950 to 1968 and were started by John L. Lewis and Harry Moses (the first president of BCOA). During this eighteen-year period, the coal industry went without a single strike over contract negotiations. This record of success in contract negotiations was hailed by John L. Lewis and the coal operators.

Mr. Judy conducted the 1968 wage negotiations for the BCOA, but clearing with the Executive Committee of BCOA. The 1968 Wage Agreement became effective October 1, 1968 and continued until September 30, 1971.

The 1968 Wage Agreement provided for the creation of a wage commission to be composed of three members which were to be named by the president of the UMWA and three were to be named by the president of the BCOA. This commission was delegated the authority to conduct a survey of wage rates in the bituminous coal mining industry. Its primary responsibility was to study the inequities in wage rates and the number of classifications that existed in the coal industry. At the completion of this survey, the commission was to report back to the president of the UMWA and the president of the BCOA their recommendations. This was to be accomplished prior to the expiration of the 1968 Wage Agreement.

The commission made its final report to the joint officers of the UMWA and BCOA sometime in October of 1971.

The commission, in its survey, found that there were approximately fourteen hundred different wage classifications in the coal industry. In some instances, it seemed like those who started coal mines would give a classification their own name, and there were as many as twenty-five different names for the same classification. The commission also found wage rate inequities in some of the same classifications that ran from $.01 to approximately $3.00 per day. In order to eliminate these inequities, the commission reduced the classifications to six pay grades for underground at deep mines, six pay grades for strip and auger mines, and five pay grades for preparation plants and other surface facilities for deep or surface mines. The commission identified the inequities in the wage rates and reduced the high rates to a certain point and increased the lower rates to the middle structure of the rates. All of these recommendations were incorporated into the 1971 Wage Agreement that became effective November 12, 1971. Both the officers of the International Union, UMW, and the officers of the BCOA were highly complimentary of the members of the Wage Commission for eliminating some of the classifications and equalizing wage rates that were later incorporated into the 1971 Wage Agreement. Following a long strike period, the 1971 agreement was concluded in New York City at the Roosevelt Hotel by the Executive Committee of BCOA and the International Officers of the UMWA.

In addition to incorporating the new job classification system, the new three-year agreement provided for (among other changes):

1. Wage increases in each of the classifications in three stages, with the top rate reaching $50 per day by the expiration of the agreement.
2. An annual protective clothing allowance for each miner of $10, increasing to $20.
3. Bereavement pay.
4. Christmas bonus.
5. Additional, graduated vacation.
6. Increase in operator royalties to the Welfare and Retirement Fund from the former 40¢ per ton to 75¢ per ton by the end of the agreement.

Also, for the first time in the history of collective bargaining in the coal industry, provisions of the many separate agreements that had been applicable heretofore, were consolidated and incorporated into one, single and basic agreement.

In September of 1968 following the 1968 negotiations and prior to those in 1971, the International Convention of the UMW was held in Denver, Colorado. This was the first time that the International Convention of the UMW was held in the western part of the United States. It has been said by some of the old-timers in the union that the main reason for holding the convention out west was to be able to give Tony Boyle more leverage in controlling the delegates to the International Convention.

Nineteen sixty-eight was also an election year. The presidential nominee of the Democratic party was Senator Hubert H. Humphrey of Minnesota, and the Republican nominee was former Vice-President Richard M. Nixon of California. Mr. Boyle extended an invitation to Senator Humphrey to speak to the delegates at the convention in Denver. Senator Humphrey accepted. Boyle and the convention endorsed Senator Humphrey's candidacy for the presidency of the United States and presented him with a check for $25,000 from Labor's Non-Partisan League, the political arm of the UMW.

This was the beginning of the encounters that Mr. Boyle would have with the problems of his leadership as president of the UMW.

Prior to Mr. Judy's presidency of BCOA in 1969, the UMWA Welfare and Pension Fund Trustees, due to lack of funds, had to cut the miners pensions from $150 per month to $115 per month. This was due to a decrease in the funding of the Welfare and Retirement Fund.

In May of 1969, the Miners for Democracy, led by Joseph Yablonski, in the UMWA, announced that Mr. Yablonski would be a candidate against W. A. Boyle for president of the UMWA in the election to be held in December of 1969.

Mr. Yablonski traveled throughout the bituminous coalfields, campaigning and criticizing the leadership of W. A. Boyle and his associates. In order for Mr. Boyle to attempt to retain the presidency of the union, he came up with the idea of raising the miners pensions back to $150 per month from $115 per month. This would lead one to believe that this move would assure Mr. Boyle's reelection by the miners. Of course, before the increase could be made, Mr. Boyle had to have a meeting of the trustees of the Welfare Fund. Mr. Boyle first called a special meeting of the International Executive Board of the UMW for Monday, June 23, 1969 to fill the vacancy created by the death of John L. Lewis as chairman and trustee of the Welfare and Retirement Fund. Mr. Lewis's death occurred on June 11, 1969, and twelve days from the time of his death, Mr. Boyle wasted no time in having himself named chairman and trustee of the Welfare Fund, succeeding Mr. Lewis.

The International Executive Board approved the appointment of W. A. Boyle as chairman and trustee of the Welfare Fund at this meeting.

Since Mr. Boyle had been named the new chairman and trustee of the Welfare Fund, he called a meeting in his office of the trustees of the fund for Tuesday, June 24, 1969. He notified George Judy that he was having a meeting of the trustees in his office. The evening before the meeting of the trustees of the Welfare Fund, Mr. Judy was forewarned of what was going to take place at the trustees meeting. The next day, he proceeded to Mr. Boyle's office for this meeting.

Mr. Boyle was present, along with the attorney for the UMW. Mr. Boyle supposedly told Mr. Judy that Josephine Roche, the other trustee of the Welfare Fund, was sick and in the hospital and could not be present for the meeting. However, Mr. Boyle said he had Miss Roche's proxy in his pocket. He stated that the purpose of the meeting was to restore the cut in the miners pensions back from $115 to $150 per month. So, Mr. Judy and Mr. Boyle agreed to raise the miners pensions at this meeting. After the meeting, Mr. Judy returned to his office, and later that afternoon, Mr. Boyle issued a press release. This release was carried throughout the coalfields. The release was, in substance, that the miners had been restored the $35 cut in their pensions to become effective August 1, 1969. This action by the trustees assured Mr. Boyle and his associates that he would have no trouble in his reelection bid against Joseph Yablonski.

On July 14, 1969, the Executive Committee of BCOA met in BCOA's offices in Washington, D.C. and terminated Mr. Judy's contract with BCOA because Mr. Judy had no authority from BCOA to go over and agree to restore the pension increase with Mr. Boyle. Mr. Judy had a three-year contract with BCOA, thus, with his contract fully paid, his employment with BCOA was terminated.

The BCOA Executive Committee proceeded and designated Guy Farmer, chief counsel for BCOA, to look after the affairs of BCOA until such time as BCOA could find a successor. Mr. Farmer's tenure was from July 14, 1969 to January 1, 1970.

In the fall of 1969, the National Coal Policy Conference (NCPC) was in the process of being dissolved because the members of this conference did not believe that the continuation of this organization was justifiable. The committee that was selected by the coal operators to dissolve the Coal Policy Conference consisted of the following: William V. Hartman, Executive Vice-President, Peabody Coal Company, St. Louis, Missouri; and P. P. Ferretti, Vice-President, Consolidation Coal Company, Pittsburgh, Pennsylvania. W. A. Boyle, president of the UMWA, was also the chairman of the NCPC.

After the NCPC was dissolved, approximately $1 million remained in the account of the NCPC. The committee decided to set up a pension of $30,000 per year for Joseph Moody, who was president of the NCPC, upon his retirement. It also set up a pension plan for the office staff who had served with him during the life of the NCPC. The remainder of these funds was donated to the UMW Welfare and Retirement Fund.

Mr. Moody was well-known in the coal industry because of his former association as president of the Southern Coal Producers Association. It was at the behest of John L. Lewis that Mr. Moody agreed to accept the presidency of the NCPC.

Because of Mr. Moody's past experience in the coal industry, the directors of BCOA decided to employ Mr. Moody as an interim president of BCOA until such time as BCOA could find a successor. His temporary employment was approved by BCOA's Board of Directors on December 12, 1969, and he was installed as president of BCOA on January 1, 1970. Since Mr. Moody had had a close relationship with W. A. Boyle while Mr. Boyle was chairman of the NCPC, and since Mr. Boyle had been instrumental in setting up Mr. Moody's pension of $30,000 per year upon his retirement, Mr. Boyle wielded undue influence with Mr. Moody.

The election campaign for the presidency of the UMWA in 1969 by Joseph A. Yablonski, the challenger, and W. A. Boyle, the incumbent, was a long and bitter one that saw both candidates touring all of the major coal areas in the United States for the miners support. This culminated on December 9, 1969 with the results of the election showing W. A. Boyle receiving 81,056 votes and Joseph A. Yablonski receiving 45,872 votes, or a plurality for W. A. Boyle of 35,184 votes.

After the election, Yablonski, his wife, Margaret, and his daughter, Charlotte, were last seen on New Year's Eve, 1969. On January 5, 1970, their bodies were discovered in the Yablonski home in Clarksville, Pennsylvania and it became apparent that they had all been murdered. The news of these murders was spread by the media throughout the country, particularly in the coalfields. The search began for the person or persons who committed these vicious crimes.

Shortly after the heinous murders of Yablonski, his wife and daughter, President Boyle announced that the UMW organization was posting a $50,000 reward for anyone who had information that would lead to the arrest and prosecution for those who had committed this crime.

In the early part of February 1970, the following persons were placed under arrest and indictments returned against them for first-degree murder by Pennsylvania authorities for the murders of Joe Yablonski, his wife, Margaret, and his daughter, Charlotte. They were Paul Eugene Gilly, Annette Gilly, Aubran Wayne "Buddy" Martin, Claude Edward Vealey and Sillous Huddleston, president of Local Union No. 3228 in LaFollette, Tennessee. Later, all of these persons were convicted, sentenced to life in prison for their roles in the murders of the Yablonskis, and are all still serving their sentences.

Prior to the murders of the Yablonskis, Joe Yablonski and his supporters filed a suit in the United States District Court in the District of Columbia asking that the election of Tony Boyle be nullified because of the many violations of the Labor Management and Disclosure Act of 1959. On June 15, 1972, upon the basis of the information that was presented to the court that Boyle had used the union's money to finance his campaign and had also used the *Mine Workers Journal* to promote his candidacy, Judge William B. Bryant found these charges to be true and ordered that the election of 1969 be declared null and void and that a new election be held from December 1-8, 1972.

This election was to be held under the supervision of the secretary of labor.

Sometime near May 29, 1972, the Miners for Democracy held a convention for coal miners in Wheeling, West Virginia for the explicit purpose of nominating a slate of candidates to run for the top three offices of the International Union, UMWA. The Miners for Democracy movement originally began as a political movement headed by Arnold Miller and a Dr. Buff who were pressuring for black lung benefits. It was expanded by Yablonski supporters following the murders of the Yablonski family. Those who were nominated are as follows:

Arnold Miller, Chairman of a seven-state Black Lung Association, Cabin Creek, WV	President
Mike Trbovich, Chairman of the Miners for Democracy of Western Pennsylvania	Vice-President
Harry Patrick, Co-chairman of the Miners for Democracy of Barracksville, WV	Secretary-Treasurer

Those placed in nomination by the UMW were:

W. A. Boyle	President
Leonard Pnakovich District 31, Fairmont, WV	Vice-President
Willbert Killion District 11, Brazille, IN	Secretary-Treasurer

When it became known to the leadership of the UMW in 1972 that Arnold Miller would be the Miners for Democracy candidate, some of the leaders in the miners union, at that point, questioned Arnold Miller's constitutional right to be a candidate for president of the UMW. After it was discussed by President Boyle and his associates, it was determined that Arnold Miller would be the easiest candidate for Boyle to defeat, and one of the main reasons for the Boyle group to consider Miller to be a weak candidate was because he was not a member in good standing with the union and was not employed in the coal industry at the time he was running for the presidency of the UMW. Also, his record of employment in the industry indicated he was in and out of the industry and did not maintain his membership in the union, as required by the UMWA constitution. Some members of District 17, UMWA, circulated Miller's work record, showing his complete record of employment in the coal industry. That record is as follows:

From 1951 to November 1959, Member of Local Union #750;
December 1959, Member of Local Union #6607;
January 1961, Member of Local Union #8843;
October 1961, Member of Local Union #6607;
November 1961, Member of Local Union #2903;
December 1961, Member of Local Union #6607;
September 1962, Member of Local Union #750;
October 1962, Member of Local Union #2903;
July 1968, Member of Local Union #750, and;
August 1970—Quit his job at Bethlehem Mines, Kayford, West Virginia and had not worked in the industry since.

It has often been said that if anyone other than Tony Boyle would have been placed on the ballot, Miller would have been easily defeated for the presidency of the union.

On June 27, 1972, the Labor Department instituted a suit against Boyle on thirteen different counts for the misuse of union funds in the national political elections of 1967, 1968 and 1969. This suit was filed in the United State District Court in the District of Columbia.

The jury found Mr. Boyle guilty on all counts, and he was sentenced to five years in a federal penitentiary and fined $130,000 in personal funds and ordered to pay to the UMWA $49,250 that had allegedly been contributed by Labor's Non-Partisan League to various candidates for public offices over a three-year period. Judge Charles Richey instructed Boyle not to "dissipate" his assets because it had been discovered during Boyle's trial that he had transferred approximately $250,000 to his wife's name to prevent using his own money to pay his fines. Pending Boyle's appeal, bond was set at more than $179,000 and Boyle was forced to spend six hours in the lockup at the U.S. District Court in the District of Columbia until his attorney could raise the bond money, which was finally posted by a surety company.

After the setbacks that Mr. Boyle had in these two court decisions, the new election of officers of the UMW was scheduled for December 1-8, 1972 under the supervision of the secretary of labor.

The results of the election, as announced by the Labor Department, appear on page 116.

The Labor Department on December 22, 1972 certified that Arnold Miller, Mike Trbovich and Harry Patrick were the duly elected officers of the UMWA, and they were sworn into office on the evening of December 22, 1972 in the United Mine Workers office in Washington, D.C., and a large demonstration in support of these new officers ensued for more than an hour.

It seemed that after Mr. Boyle was ousted as president of the UMW, his problems began to mount. He attempted suicide on September 23, 1973, the day he was to surrender to begin his sentence for using union money for political donations.

William Turnblazer, president of District 19, Albert Pass, secretary-treasurer of District 19 and William Prater, a representative of District 19, were all indicted as accessories to the murders of Joe Yablonski, his wife and daughter. All of these individuals were close associates of Mr. Boyle.

After these individuals were tried and convicted for their

DECEMBER 1-8, 1972 UWMA OFFICERS ELECTION RESULTS

District	Miller	Boyle	Trbovich	Pnakovich	Patrick	Killion
2	5,465	5,396	5,363	5,115	5,179	5,295
3	217	522	215	483	222	480
4	2,967	2,184	2,937	2,093	2,954	2,044
5	5,567	3,315	5,408	3,237	5,465	3,177
6	6,384	2,112	6,079	2,290	6,325	2,053
10	84	152	81	151	85	150
11	1,089	2,157	1,028	2,167	793	2,424
14	410	514	506	506	398	515
15	935	912	873	946	918	904
17	7,583	4,853	7,086	5,029	7,693	4,444
19	594	2,454	569	2,343	567	2,341
20	1,444	4,170	1,109	4,081	1,164	4,039
21	135	574	138	567	143	560
22	909	1,251	896	1,246	912	1,237
23	2,204	3,376	2,131	3,257	2,136	3,265
25	9,170	1,845	8,854	1,878	8,880	1,825
27	123	273	124	269	128	264
28	4,127	2,913	3,895	2,888	4,037	2,732
29	6,544	6,566	6,086	6,640	6,597	6,126
30	2,144	3,374	2,035	3,315	2,124	3,206
31	5,681	3,343	5,120	3,758	5,636	3,231
Totals	70,373	56,334	66,910	56,440	68,696	54,645
Plurality of Victors . . .	14,039		10,470		14,051	

SOURCE: United Mine Workers Journal, December 1971, p. 3.

part in the terrible murders of the Yablonskis, they were all sentenced to life in prison.

In 1973, Boyle also was convicted for ordering the murder of Joe Yablonski. He was sentenced to three life terms in 1974 for first-degree murder and served time at the infirmary of Western State Penitentiary at Pittsburgh, Pennsylvania until his conviction was overturned in January of 1977 by the Pennsylvania Supreme Court because Boyle hadn't been allowed to present a full defense. A new trial was ordered for Boyle.

At Boyle's second trial in 1978, Albert Pass testified that Boyle had ordered the assassination of Joe Yablonski and had provided $20,000 for hired assassins. On February 17, 1978, Boyle was found guilty on three counts of first-degree murder and was given a life sentence on each count. Boyle died in a Pennsylvania prison on May 31, 1986 at the age of eighty-three. Upon his death, his body was cremated.

CHAPTER XXIV

Miners for Democracy

After Arnold Miller was sworn into office, his first official act was to revoke all of the provisional districts under the jurisdiction of the United Mine Workers of America and grant them the authority to establish district autonomy, meaning they could elect their own district officials. He followed this by dismissing all of the representatives of the UMWA and cutting the salaries of the legal department to such an extent that they all resigned their positions, with the exception of Harrison Combs and Willard Owens, and possibly two or three others who took the salary cuts and continued to hold on to their positions.

President Miller began to fill many of these vacancies with new, inexperienced people and several young lawyers. He named Joseph A. "Chip" Yablonski as the new chief counsel for the UMWA. Yablonski was also named as a director of the National Bank of Washington from which he received remuneration for his services in addition to his salary of more than $50,000 per year as chief counsel of the UMW.

In the course of the new leadership of the union, it was charged that Chip Yablonski had taken advantage of the union. In addition to his salary as chief counsel and a member of the National Bank of Washington Board of Directors, he was also charging the union for the legal services he was performing for the UMWA. When the International Executive Board became aware of the advantage that Chip Yablonski was taking of the union, they terminated his services. Harrison Combs, formerly in charge of the legal department under Tony Boyle, was named chief counsel succeeding Yablonski. He was later named as one of the trustees of the UMW Welfare and Retirement Fund representing the UMW.

In 1971, the famous *Blankenship vs. Boyle* breach of trust,

Case No. 2186, a class action suit, was filed in the United States District Court in the District of Columbia.

The substance of this suit was that there were many miners in the coal industry who had been denied their pension because they had not been employed in the industry on May 29, 1946 when the provision, creating the Welfare and Retirement Fund, was written into the national agreement. Also, the trustees of the Welfare and Retirement Fund were not being selected in accordance with federal law, and there were thousands of coal miners who had been denied credit toward their pension at the time of their retirement because they had gone to work in a nonunion operation, and the union was using this practice as a means to deny these miners their pensions.

On April 28, 1971, the federal court ruled in behalf of the plantiff. This decision affected about sixteen thousand miners who were therefore automatically eligible to receive a pension from the UMW Welfare and Retirement Fund. This decision further provided for the reorganization of the trustees of the Welfare Fund and prohibited the president of the international union from being a trustee of the fund, or a director of the National Bank of Washington.

This decision renewed the membership of the miners affected by the case in the UMW and made them eligible to vote in the election in 1972 for the presidency between W. A. Boyle and Arnold Miller. This decision was also a contributing factor in influencing many coal miners throughout the coal industry to vote for the officers of the Miners for Democracy, who were later declared the duly elected officers of the UMW by the Labor Department.

During the course of this litigation, the Black Lung Association of West Virginia, under the leadership of the same persons who participated in the Blankenship case, was instrumental in having the Legislature of West Virginia pass a black lung law making a miner eligible to receive black lung benefits. This, too, was a contributing factor in the election of the new officers who ran under the auspices of the Miners for Democracy.

Arnold Miller, the new president; Mike Trbovich, the new vice-president; and Harry Patrick, the newly elected secretary-treasurer, began to take charge of the affairs of the International Union, UMWA, immediately after they were sworn into office, and they had a free hand in designating new personnel to operate the UMW organization.

These men were assisted in their bid to capture the leadership of the UMW by many outside organizations, and funds were made available through several family foundations under the influence of Joseph L. Rauh, a noted liberal Washington attorney who assisted in advising the Miners for Democracy in helping defeat the Boyle slate, thus, beginning a new era in the UMW leadership.

"Arnold Ray Miller was born in 1923 in the mining village of Leewood, a few miles up Cabin Creek from the Kanawha River, and about twenty miles from Charleston. He was raised by his mother's parents, his father and mother having separated before he was born.

"He left school after finishing the seventh grade to go to work in the coal mines, the sole industry on Cabin Creek.

"Arnold enlisted early in the U. S. Army in World War II. He was sent overseas to the European theatre where he was severely wounded. During two years of hospitalization he was operated on nineteen times. The only remaining evidence of these operations was a disfigured ear.

"After his discharge from the Army, Arnold came back to Cabin Creek and resumed working in the coal mines. He later operated a commercial garage and apparently attained considerable skill as an auto mechanic. After a short period at the garage, he went back to work in the mines and continued there (intermittently) until 1970 when he became disabled with black lung and possibly other related diseases. He was a heavy cigarette smoker.

"Miller understandably took an active interest in the new Black Lung Association and, after a brief action-filled period, was elected to head the local group.

"Miller spent seven sometimes strange years as president of the UMW. They were unusual years for the Nation. They included the Arab Oil Embargo and the super inflation of which the Embargo was a partial cause. In that short period, crude oil prices increased by about ten to one, and coal prices by about half that much. Under extreme circumstances, the best grades of West Virginia coal sold briefly for as much as $100 per ton. Overnight, coal millionaires became commonplace. Coal equipment salesmen built million dollar homes. However, this over-inflation of coal prices didn't last long.

"During Miller's tenure, there were frequent changes in the ownership of large coal mining complexes due to their acquisitions by "Big Oil" and other conglomerates. The purchases were apparently made as hedges against ultimate oil depletion.

"From the beginning of his first term in office, Miller had organizational and administrative difficulties. Neither he, nor his vice president Mike Trbovich, nor his secretary-treasurer, Harry Patrick, had had previous administrative or executive experience.

"A leader is an individual who sees the problems around him more clearly than others, and is able to communicate his ideas to others so vividly that they will join him in forceful action. Leadership ability was something Arnold Miller . . . simply didn't have. Granted this disability, the Miller administration accomplished a great deal. Some say it was despite its president.

"In his efforts to bring back halcyon days to the UMW, Miller relied on implementing the battle cry of the MFD during the election campaign: Democracy for All. He and his cohorts did, in fact, establish rank and file approval of all contracts with the BCOA. The Miller Administration also established an in-between group whose consent was necessary to validate contracts. This requirement turned out to be an embarassment.

"Miller weakened his chance for re-election by not trying to influence the election of his friends to the Executive Board. He failed to understand that democracies, unguided, sometimes have the characteristics of irresponsible mobs. He may have relied too much on the educated liberal young men in the organization. They became easy targets for the old hands who charged them with being Communists or Socialists, which in most cases, was untrue.

"After a short period in office, Miller lost the confidence and services of his vice-president, Mike Trbovich, who may still have resented having lost to Miller for the UMW presidency. Miller's secretary-treasurer, Harry Patrick, also departed and opposed Miller for a second term.

"Miller . . . didn't get along well with the BCOA. He didn't trust them and was unable to hold his own, man to man, with their spokesperson in contract negotiations. Nevertheless, the first contract negotiated with the BCOA, during his tenure, gave the UMW several advantages.

"In the UMW Constitutional Convention, several desired constitutional changes were made. UMW Districts were given back their long-lost autonomy.

"Despite considerable progress in carrying out MFD plans, Miller failed to win the approbation of several MFD leaders.

"Wildcat strikes which UMW president Lewis had put down with an iron hand when they didn't suit his purposes, plagued the Miller administration and, of course, didn't help its standing with the BCOA.

"One of the reasons for the damage done by wildcat strikes was the fetish of the picket line. Almost everyone could set up a picket line for little or no reason and it was considered sacrosanct regardless of the damage to innocent parties. Under these circumstances some picket lines would have been laughable if they hadn't had serious consequences.

"There were many wildcat strikes in the Kanawha area in the Miller era. They persisted notwithstanding Court rulings that they violated the Taft-Hartley Act when they were contrary to the terms of a valid contract. (This situation was of immense frustration to the coal operators, affecting their profitability and customer relations.)

"In one way or another, Arnold Miller not only survived his first five-year term, he also, to the surprise of many, announced his candidacy for a second term. At the time of this announcement his vice-president was Sam Church, a miner from western Virginia. His secretary-treasurer was Willard Esselstyn from Pennsylvania.

"Miller's opponents in his second try for the UMW Presidency were Lee Roy Patterson, a strip miner from Kentucky, a long-time Miller critic, and Harry Patrick from northern West Virginia, who had been Miller's secretary-treasurer.

"In the 1976 UMW election, Miller received 55,000 votes, Patterson, 49,000 and Patrick, 34,500. The fact that Miller failed to win a majority of the votes cast was not necessarily disabling.

"During the first part of Miller's second term, decreased coal tonnage made inevitable an unpopular cutback in welfare and retirement payments.

"1977 was the year for a contract between the UMW and the BCOA. An important and unresolved issue was the extent of penalties for unauthorized strikes. A Communist dominated organization, the "Miners' Right to Strike Committee," helped keep the issue from being objectively resolved.

"Another issue was the amount of contributions by the BCOA into the UWM Welfare and Retirement Fund. Severe inflation seemed to call for higher cash benefits.

"When, on December 6, 1977, no contract had been signed, the miners, as was their custom, walked out. The strike quickly became controversial among President Carter's administrators. Some of his advisors favored the UMW; some the BCOA. President Carter wavered in between.

"On February 6, 1978, the strike negotiators announced they had reached an agreement. President Miller, for health reasons, had been absent from most of the negotiations. In the agreement, in addition to a 37% wage increase over a three year period, the miners got most of the other benefits they asked for. Unfortunately, however, that was not the final word.

"The official bargaining council which the Miller administration had established rejected the agreement by a five to one vote and the strike continued. The council's action sent the negotiators back to the bargaining table. The council then proposed a substitute agreement which was rejected two to one by rank-and-file miners. These rejections by two different bodies of the same organization left matters in a state of confusion.

"In an attempt to settle the strike, President Carter, on March 6, 1978, invoked the (national emergency procedure of the) Taft-Hartley Act. It required, as a second step, that an appointed commission decide whether the coal strike posed a real and present danger to the U.S. economy. The question was complicated by the fact that some large plants had accumulated huge stockpiles of coal in anticipation of a strike. Furthermore, about half the Nation's coal producing capacity was non-union.

"The commission ruled that no national emergency existed, and the Carter administration asked the BCOA and the UMW to resume bargaining. A new BCOA offer which raised wages only slightly over former offers but increased welfare and retirement payments considerably was approved by the miners on March 26, 1978.

"During this long and disruptive strike, President Carter appointed a coal commission to determine, on a long-range basis, what ailed the strike-prone coal industry. He named West Virginia's Democratic Governor, Jay Rockefeller, as chairman of the commission.

"The Rockefeller commission worked long and hard trying to solve the national coal riddle but without spectacular results. Its principal recommendations were centered around ways and means to increase coal production and to improve coal markets. One of its specific recommendations was to require, by law, the use of coal at all electric power plants. These plants were the largest consumers of U. S. coal. The commission offered no remedy for the constant friction between operators and miners. Neither did anyone else.

"While the long coal strike was in progress President Miller's influence in the UMW declined to near zero. For most purposes, by 1978, his vice-president, Sam Church, ran the show. Miller's situation became so bad that a petition for his recall got 31,000 signatures. To nobody's surprise, Arnold Miller, after seven years as UMW president, resigned and Sam Church became president."[1]

When Mr. Lewis retired in January of 1960, he was convinced that the union had reached its peak, as the total assets of the union as shown on Form LM-2 of the Department of Labor indicated total assets of $110,315,080. Included in the assets, was the stock in the National Bank of Washington and other real estate property that was worth several million dollars. The total number of working coal miners was approximately 172,000 and the prestige and influence of the United Mine Workers was recognized throughout the coal industry, in labor and government circles. Mr. Lewis's successors, Thomas Kennedy and W. A. Boyle, followed his example in the administration of the affairs of the union. When Mr. Boyle relinquished his office, the financial statement, Form LM-2, for the year 1972, dated May 10, 1973, showed total assets of $83,245,828. This period covered the last year of Boyle's leadership. At this point, the report showed there were 147 employees attached to the international office.

In the first year of Miller's leadership, the international office had grown to 152 employees. At the end of 1973, the total assets of the UMW were $83,441,998.

As the administration of President Arnold Miller continued, the assets, prestige, and influence of the union began to drop. From the period of 1973 to May of 1980, the assets of the International Union dropped to $39,233,411, a decrease from 1973 of $44,208,587, or a 52.9% drop in the assets of the union under the administration of Arnold Miller.

During this period, Mr. Miller attempted to sell all of the stock of the UMW in the National Bank of Washington to a group of buyers. When he submitted his proposal to the International Executive Board, it was rejected. In 1977, he sold the building located on 15th and K streets for $966,959. In the meantime, the union dues under his administration increased to $12 per month.

Most of the increased expenditure under his administration was due to increasing the international staff from a total of 147

1. V. B. Harris, *Kanawha's Black Gold and the Miners' Rebellions* (unpublished manuscript), Chapter Nine, Arnold Miller, UMW President 1972-1979.

in 1972, under the last year of former President W. A. Boyle, to an increase by Arnold Miller in 1979 to 280. Many jobs and new titles were initiated during this period that were not necessary. This does not include the additional jobs created in the various districts under the jurisdiction of the UMWA. Most of the districts increased their staffs, not because it was necessary for the benefit of the coal miner, but because it was politically expedient for those who ran for international and district offices. This was one creature that was developed under the guise of the democracy being instituted in the UMW international and district organizations.

During Arnold Miller's regime, the coal industry suffered its worst siege of illegal work stoppages in the history of the bituminous coal industry. Many of these strikes were black lung strikes, and in most instances, the grievance procedure was not followed as provided for in the national agreement. This irresponsible leadership cost the coal industry millions of doilars in legal fees seeking court orders to force the union to return its members back to work and did not include the loss of millions of tons of coal production due to these illegal strikes.

The relationship between mine management and the union dropped to its lowest ebb in the coal industry. President Miller, with approximately one year as head of the UMW, held his first constitutional convention in Pittsburgh, Pennsylvania, in December of 1973, and many changes took place in the constitution of the UMW. One of the major changes was the elimination of the old wage-policy committee to approve negotiated agreements that had been so successful under the leadership of John L. Lewis. Arnold Miller instituted a new amendment to the constitution called the bargaining council. The bargaining council consisted of the international officers, the district officers and members elected from the mines. They were given the authority to accept or reject any agreement made between the coal industry and the union. Any acceptance of a negotiated agreement, under the new system, further required ratification by a majority vote of the UMWA members.

This new amendment eliminated the contract-approval authority that the international officers had under the policy committee and placed it in the hands of the bargaining council. The representatives of the BCOA had to submit the agreements made between the bargaining council and the BCOA negoatiators to a full committee of the bargaining council before any final agreement could be made. And then, the further step required ratification by the full membership.

The leadership of the BCOA, its chairman, N. T. Camicia, who presided from April 4, 1972 to April 1976, and some of BCOA's directors, catered to Arnold Miller and went to his office and congratulated him upon his election as the new president of the UMW with the hope that Mr. Miller would give them some special consideration and develop some discipline in the union to combat the rampant wildcat strikes. Instead, Miller did just the opposite. He encouraged wildcat strikes and took no disciplinary action against the local union officers who promoted these illegal strikes throughout the coal industry, and he would not respond to any messages from the industry advising him of these illegal work stoppages.

Prior to the UMWA convention in December of 1973, Miller discussed with Mr. Camicia changing the grievance procedure of the National Wage Agreement. Mr. Camicia appointed a committee from BCOA to meet with a committee designated by Mr. Miller to study the grievance procedure and make a recommendation. In the meantime, BCOA's committee agreed to push for a change in the grievance procedure. The change involved a complicated arrangement of arbitration panels throughout the industry, and an appellate body in the form of the "chief umpire" to take appeals from the rulings of panel arbitrators. The proposal grew out of a joint BCOA-union study provided for in the 1971 agreement. BCOA's committee was dispatched to Pittsburgh to the Mine Workers convention to discuss the proposed changes in the grievance procedure with Mr. Miller, and Mr. Miller agreed to the changes. He submitted the changes to the convention for their approval, and after considerable discussion by the delegates to the convention, the convention voted to reject the entire proposal.

After the convention, the bargaining council and the members of the BCOA negotiating committee started holding their meetings, and the grievance procedure was brought up again by the negotiators for BCOA. Mr. Camicia told the BCOA negotiating committee to agree to the changes in the procedure, because in his opinion these changes would not cost the industry any money and would tend to alleviate the wildcat strike situation. The new grievance procedure was written into the 1974 contract and became effective December 6, 1974. This was the same grievance procedure that had been rejected a year earlier by the Mine Workers' convention. As a result of incorporating this new grievance procedure into the contract, it has *proliferated grievance handling* and cost the industry millions and millions of dollars.

This new grievance procedure was an elaboration of a system that had been discarded by Mr. Lewis many years before. Mr. Lewis always expressed that the arbitration procedure was nothing but a "buckpassing apparatus" used by the Mine Workers and some representatives of the coal industry. It was a method that discouraged both parties from settling any grievance on the local level—it was always passed off to an arbitrator who knew nothing about the conditions of the coal industry or the operation of a coal mine. It was Mr. Lewis's thought that this was an unfair way to settle mine grievances. The old agreement provided that after the representatives of the union, and the mine committee, and the operators could not settle their grievance, it was then submitted to a joint board which consisted of two members of the union and two members of the mine management. They would make a final decision. Mr. Lewis thought this was the best method of settling grievances because both members of the board representing the mine workers and both members representing the operator were most familiar with the conditions in the operation of a coal mine rather than an outsider who knew nothing about them. They were better able to settle grievances more amicably. Arbitration to an impartial umpire was a last resort, and hopefully, its use would be minimized. The 1974 agreement not only acknowledged arbitration as the final step in the grievance procedure, but engrafted on it additional involvement by outsiders in the form of panels of arbitrators, an Arbitration Review Board, and a "chief umpire."

It is significant to note that on April 4, 1972, Mr. N. T. Camicia was selected to be the new chairman of the BCOA succeeding Heath Larry, vice-chairman of the board of United States Steel. Also, the Miners for Democracy took over the leadership of the UMWA on December 22, 1972.

The leadership of the coal industry was convinced, beyond any doubt, that the new leadership of the union was very weak, and Mr. Camicia, the new chairman of BCOA, thought the best strategy for the industry to pursue against this weak leadership was to go along with the union's leadership, having in mind that they could extract from the union more concessions to benefit the industry. He was able to convince Consolidation Coal Company, Peabody Coal Company, and U. S. Steel that it was best to get along with the new union leadership. Since the bylaws of BCOA provided that members of BCOA would vote on a tonnage basis on any issue that was being questioned by the other member companies, it developed that the three companies mentioned above had, by far, enough tonnage to force the other members of BCOA to go along with whatever course they elected to pursue with the union.

Shortly after Arnold Miller's election as president of the union, Mr. Camicia, as BCOA chairman, began having meetings between himself and Mr. Miller. Mr. Camicia felt that the best way to deal with the union was to make concessions to them which would result in a more responsible union and good future relationships, with benefits to BCOA.

Miller was also successful in getting the check-off provision in the contract amended to give the union authority to institute special assessments upon the mine workers so they could check-off for a credit union that was later established by the union. Also, this change in the check-off in the national agreement provided the right for a check-off for Coal Miners Political Action Committee (COMPAC). This was the lobbying agency of the UMW.

As a result of these changes, the union began to assess the miners to raise enough money to establish a credit union and to expand COMPAC's lobbying agencies in all of the coal producing state legislatures and the United States Congress. This was the first time in the coal industry that coal operators were required to check-off for COMPAC and turn this money over to the union to be used for lobbying throughout the coal producing state legislatures and the Congress of the United States to pass laws against the industry. This money is also used for PACs (political action committees) for the purpose of helping to elect candidates who are favorable to the union. It is a sad and costly commentary when the coal industry indirectly provides the funds for the union only to have legislation passed against the industry. The coal industry no longer wonders why these laws are enacted.

On January 1, 1974, under Mr. Camicia's leadership, the BCOA went outside the coal industry and employed Walter Wallace as the new president of BCOA. Mr. Wallace was a lawyer and had also served as assistant secretary of labor in the Eisenhower Administration. At the time of his employment, he was the director of personnel and labor relations for the Hudson Paper Company in New York state. Mr. Wallace was given a five-year contract at $75,000 per year, an automobile, and membership in the Executive Club and a few other clubs in the city of Washington. Mr. Wallace had no knowledge or experience in the coal industry. Mr. Wallace's tenure as president of BCOA was of short duration. At the urging of BCOA members, he was personally dismissed by Chairman Camicia on April 8, 1975. BCOA was then required to pay off Wallace's three year and nine month contract which amounted to approximately $300,000. Mr. Wallace was not given any advance notice of his dismissal. He was to have chaired the annual meeting of BCOA on April 9, 1975. The night before, Mr.

Camicia went to Mr. Wallace's office and terminated his services. On the next day, Mr. Camicia announced at the annual meeting that Mr. Joseph P. Brennan had been employed as the new president of BCOA. Mr. Brennan had been formerly associated with the UMW under the W. A. "Tony" Boyle administration. When Arnold Miller was sworn into office as president of the union, Brennan was one of those employees who had been dismissed. Brennan had been serving as director of research in the UMW. After his tenure with the union, he was employed by the National Coal Association in National Coal's economics department. Although Brennan has continued to serve as president of BCOA ever since his employment, his activities as president of BCOA have been very limited, and he has not participated in any of the wage negotiations. Mr. Camicia announced to the directors on April 9, 1975 that Mr. Brennan had been employed at a salary of $30,000 per year and that BCOA was not required to sign a contract for his employment.

After Joe Brennan's employment, it became apparent among the operators in BCOA that Brennan was only a messenger for Mr. Camicia. Almost daily, he went down to Arnold Miller's office and then back to his own office to relay Mr. Miller's message to Mr. Camicia.

At the time Camicia served as chairman of BCOA, he was the president and CEO of the Pittston Company which had a number of subsidiary companies in Virginia, West Virginia and Kentucky. He came to Pittston from the General Dynamics coal operations in Illinois where he was the vice-president of operations. Most of his mining career had theretofore been with the Island Creek Coal Company and its related Pond Creek Pocahontas Coal Company, where he had risen to vice-president of operations. His mentor was Raymond Salvati, long-time top official in the company and was well-known throughout the industry. Camicia was a mining engineer, having received his degree from Virginia Polytechnic Institute. He was raised in the coalfields of McDowell County, West Virginia.

Mr. Camicia, prior to being promoted to vice-president of operations of Island Creek, was division superintendent of the Mud Fork Division at Holden, West Virginia. He was involved in a confrontation with a union "mule train" that organized in Mingo County sometime in 1950. This mule train went to the No. 7 mine of Island Creek at Holden, West Virginia. The term "mule train" applies to a group of radicals who came to No. 7 mine in a caravan to picket and stop the employees from working.

The company had been aware of the efforts of pickets to shut down mines around Logan and knew that John B. Reed and Bill Maynard, local union officials at Island Creek's mine near Delbarton, were very active in this effort.

On the day in question, Nick Camicia received word that a group of pickets had formed a "mule train" in Delbarton and that they were heading towards Logan County via No. 22 Mountain and Holden. The company didn't know whether they would hit the No. 1 mine, No. 7 mine or go towards Logan.

Dutch Barber and one other deputy sheriff sent word to the company that they were in a car ahead of the union convoy. The company had also been told that the pickets had a large number of cars and had intimidated the merchants in Delbarton to give them money for gasoline and liquor.

Approximately two hours after receiving this information, Nick Camicia, division superintendent; Ralph Thompson, the mine superintendent; and several other Island Creek officials, were in Mr. Thompson's office at the No. 7 mine when they heard a loudspeaker coming towards the No. 7 mine playing "Mule Train" and "Sixteen Tons."

A large caravan of cars pulled up at the mine on the road between the tipple and the mine office. The second car in the caravan had the loudspeaker, and men were hanging on the outside of some of the cars. Some of these men had weapons—pistols, rifles or shotguns.

A group broke into the mine office where the company officials were and immediately ordered them to leave. They started to walk out between lines of the men gathered in the office. As Nick Camicia started to leave, he objected to being pushed and said something in protest to the pickets. Just as he reached the doorway to the outside, someone hit him and knocked him to the ground outside. Several others formed a ring around him and started kicking him.

Another company official, who had been in front of Mr. Camicia as they were leaving the mine office, turned around and stated that he thought Mr. Camicia had had enough and they should let him up. The company official recognized none of these men. They were not Island Creek men. It was later determined that the men directly involved at the mine office were probably from Landau-Puritan Mines. One of them, the one who first hit Nick Camicia, was later identified as a no-good who at times drove a taxi in Delbarton, probably not a miner at all.

In any event, the group who formed the circle around Nick

Camicia let him up. He was led to a company car so that he could be taken to the Holden Hospital. The car was parked a short distance from the mine office, but was blocked by the "mule train" cars. One man was finally recognized as Hubert Watson, a young ex-marine and mine committeemen at No. 26 Mine at Rockhouse. He was enlisted to get people to move some of the cars so that Nick Camicia could be taken to the hospital. Mr. Camicia insisted on driving himself to the hospital.

Although a medical examination revealed that Mr. Camicia had suffered a concussion and contusions, they confirmed that he suffered no serious injuries.

At the mine office after Camicia had been taken to the hospital, the "mule train" members still filled the office, and it was impossible to get to the only door leading to the exterior of the mine office building. They pushed and shoved Ralph Thompson, Ed Walsh, Harrison Maynard and Rex Browning into the superintendent's office.

John B. Reed started a conversation by saying that they wanted the bosses removed from the mine, as they were taking the place of the union pumpers. Ralph Thompson told them that it had always been customary to leave the union pumpers on duty when the contract expired; and when a new contract was signed, the union pumpers would receive the extra benefits retroactive to the time the old contract expired. This time, the UMWA had pulled the union pumpers from the mine. Since No. 7 mine was a large mine, extending nine miles underground, with a large number of pumps and twenty-seven union pumpers on the payroll, if the pumps were not serviced and looked after, the whole mine would be flooded, and it would take a year or so to put the mine back into operation.

Ralph Thompson told John B. Reed if they let the pumpers return as they had worked in the past, then the bosses would not do the pumping of water; if the union did not let the union pumpers return to work, then, the foremen at the mine would continue to operate the pumps.

Before anything more could be said, Bill Maynard made a statement concerning mistreatment of union men at the No. 7 mine. Bill Maynard asked who Ed Walsh was. Ed Walsh replied, "I am Ed Walsh." Ed Walsh had a mine accident and almost lost both his legs. He had been off from his regular job and was second shift foreman, a position he had held for two years, and he had helped Rex Browning as timekeeper in the mine office. The union men started hitting, pushing and kicking Ed Walsh and finally took him through the crowd to the outside. He received severe physical damage from these men.

Joe Walsh, Ed Walsh's son, lived near the mine office, and he came out of his house and helped rescue his father.

Harrison Maynard, tipple boss at No. 7 mine, was also led out of the mine office. Harrison was a preacher and had some friends in the crowd. His friends helped him through the crowd and then one of his friends screamed "Run, preacher, run!" Harrison Maynard and Ed Walsh lived just about three hundred yards from the mine office.

Only Ralph Thompson and Rex Browning remained in the mine office. The pickets decided to make both of them leave by shoving, kicking and slugging them from the office building to the garage where Ralph Thompson kept his car. Both men were forced into Ralph Thompson's car, but Thompson could not back the car out of the garage. The union men filled the garage and were in the rear of the car so that he could not back the car out without hurting several of the union men. These men were asked to clear the road so the car could be backed out. Finally, they moved, and Thompson was able to get the car out of the garage, but they still surrounded the car. It was then that John B. Reed and other union men remembered that they had not completed the job they had planned for running the "mule train" which was to remove the foremen from the mine, shut down the pumps, and let the mine flood. The union men pushed Thompson's car back into the garage and made both Thompson and Browning get out of the car. Thompson argued with them, giving them several reasons why the pumps had to operate. They insisted on going inside the mine and removing the foremen themselves. Thompson told them that it was against the West Virginia state law for drunks to enter a mine. After much discussion, Thompson agreed to let a motorman, Quincy Muncy, who worked at No. 7 mine, go with him inside the mine to bring the foreman out.

At this time, one of the drunks said "Rex Browning has a gun." Several of the union men began to beat up on Browning. After the beating started, Stanley Drozell, Mitchell Farris and Walter Hunter, the mine committee at No. 7 mine, came to Browning's rescue. Walter Hunter, who was a black man, told the union men that Browning was the timekeeper and that they should not bother him. They told Walter Hunter to keep his mouth shut or they would beat him also.

These men finally managed to get Browning to the edge of the crowd. (It was estimated that 250 union men participated in the "mule train.") Browning was told by the No. 7 mine committee to run for his life. He ran from the mine office to Sycamore Camp, jumping fences until he arrived at Tony Dardi's house. Mr. Dardi hid Browning in one of his coat closets, and when they knocked on the door of Mr. Dardi's house looking for Browning, Mr. Dardi told them that he had not seen Browning and they departed. Clinton "Buddy" Ferguson finally found out where Browning was and came to the Dardi house and picked up Browning and took him back to his home in Logan. After this incident, Browning was black and blue and suffered tremendous pain for two weeks.

Later at the mine office, Sgt. R. D. Platt of the state police arrived and escorted the foremen to their homes. It was later said that Ralph Thompson was going to take the foremen out at No. 1 mine, as No. 7 mine and No. 1 mine were connected underground. Sergeant Platt, at the time, said to bring the foreman out at No. 7 mine. They came out at No. 7 mine, and the foremen were not harmed by the members of the union "mule train."

Mr. Camicia remained in the hospital, and the company official who had accompanied him there returned to the mine. By then, Sergeant Platt was there, alone, talking to the strikers. A group had Ralph Thompson hostage, insisting that he accompany them into the mine to where the foremen on duty that day were located. Mr. Thompson refused, and finally, through the intervention of Sergeant Platt, the group let Mr. Thompson go. The crowd had grown by that time. Besides the men in the "mule train," some Island Creek men, for example, Jim Neece, president of the local union at No. 7 mine and Walter Hunter, a member of the mine committee had joined the crowd. But, they said they had nothing to do with the affair.

After a while, Sergeant Platt convinced the leaders that they could only legally stay on the highway and should leave the company property.

The group slowly disbanded.

This incident resulted in Island Creek obtaining a temporary injunction which was later made permanent. This same injunction was in effect when John B. Reed, Bill Maynard, and a bunch of men from the No. 24 mine beat up the foremen the day the company had a mass lay-off. This resulted in contempt of court and criminal convictions.

Both of these statements are recollections of two of the participants who were molested by this group of radicals. Our purpose in relating this incident is to wonder why Mr. Camicia became so conciliatory to the union. Is it that, as a result of this incident, Mr. Camicia tried to redeem himself through his contracts and meetings with Arnold Miller while he was chairman of BCOA? Or was he expressing the desires of some of his BCOA members? Camicia continued to make concessions to Arnold Miller throughout his term as chairman of BCOA.

During all of his employment with Island Creek, Mr. Salvati (more or less) let Camicia have his own way, and there were several men employed by Island Creek, that were very good coal men, but Camicia forced them to leave.

In 1951, James L. Hamilton went to work as Executive Vice-President of Island Creek Coal Company. Mr. Hamilton was the first outsider to be employed by Island Creek in such a rank. This did not set well with Mr. Camicia and other Island Creek officials. Hamilton became Mr. Camicia's immediate supervisor. Hamilton was formerly employed by Republic Steel Corporation as president of its coal division with headquarters in Uniontown, Pennsylvania.

Mr. Hamilton continued as Executive Vice-President until November of 1964, when he succeeded Mr. Salvati as chairman of the board and Chief Executive Officer. This change, in the direction of the operation of Island Creek Coal Company, was influenced by it's former chairman of the board, Cyrus Eaton, who was an outstanding industrialist from Cleveland, Ohio, the chairman of the board of the Cleveland Trust Company, one of the largest banking institutions in that area, and also, chairman of the board of the Otis Elevator Company. Through Mr. Eaton's close friendship with John L. Lewis, he continued to play an active role as a director of Island Creek and was the most influential member of the directors of Island Creek. This influence gave James L. Hamilton unlimited influence and authority in formulating the future policies of Island Creek.

On November 2, 1964, when it was announced by Island Creek Coal Company that Mr. Hamilton had been elected chairman of the board and CEO of Island Creek, it was also announced that William A. Bellano, a former employee of Island Creek, had been employed as the new president of Island Creek Coal Company, effective January 1, 1965.

Mr. Bellano, at that time, had nearly thirty years experience in the mining industry in the United States and South America in engineering, operations, sales and administration. Mr. Bellano had been president of the Glen Alden Coal Company in the anthracite region for a number of years. Mr. Bellano was also formerly the president of the Gulf Sulphur Company and at one time he had been in the operating department and later became vice-president of engineering of the International Minerals and Chemicals Corporation. Mr. Bellano was a graduate of the mining engineering school of Pennsylvania State University.

Mr. Camicia was unaware of Island Creek's announcement of November 2, 1964. He read the announcement in the newspaper in his office the following day that Bellano had been named the new president of Island Creek. After reading the announcement, Mr. Camicia, as well as his secretary, picked up their personal belongings and he resigned as Executive Vice-President of Island Creek Coal Company, effective November 3, 1964. In February of 1965, a few months after his resignation, Mr. Camicia was employed through the influence of Mr. Salvati and the president of the Freeman Coal Company of Chicago, Illinois, as Executive Vice-President of Freeman Coal Company. Freeman Coal Company was a subsidiary of General Dynamics Corporation. After Camicia had been with Freeman Coal Company for more than three years, the president of Freeman Coal Company accepted a higher executive position with General Dynamics, and Mr. Camicia was promoted to president. In 1969, through the influence again of Mr. Salvati, Mr. Camicia was employed as the president of the Pittston Company by J. P. Routh, chairman and CEO of the Pittston Company, headquartered in New York. You will note, that throughout Mr. Camicia's career, all of the positions he received were through the influence of Raymond E. Salvati, who was an outstanding individual and wielded much influence in the coal industry, particularly in the National Coal Association; the American Mining Congress and Southern Coal Producers Association. Upon the retirement of J. P. Routh of the Pittston Company in 1976, Mr. Camicia was promoted to the position of chairman of the board and CEO of the Pittston Company.

It is well to note that Mr. Camicia, when named vice-president of operations for Island Creek, William Bellano was general manager of the Marianna Division of Island Creek in Wyoming County, West Virginia, and one of Mr. Camicia's official acts after becoming vice-president of operations was to discharge Mr. Bellano from his employment with Island Creek. Many of the supervisory personnel of Island Creek commented that Mr. Camicia knew that when Mr. Bellano assumed his new position as president of Island Creek Coal Company, he would have dismissed Mr. Camicia because there never were any good feelings between these two individuals when Bellano had been formerly employed with Island Creek. After Mr. Camicia became chairman and CEO of Pittston, he became very active in BCOA. Many of those who knew him while he was at Island Creek felt that he was a very weak individual. From December 22, 1972 to April 8, 1973, it became apparent that the leadership in the coal industry, both from the union standpoint and that of BCOA, was very weak. With a new regime in the leadership of the miners union, this was the first time in the history of the miners union that its new leadership was inexperienced and had no prior knowledge of the workings of the union and was completely ignorant of handling labor relations in the coal industry. With all of these new people to work with, it would lead one to believe that this was an ideal situation for the coal operators to find themselves in. They must be firm and provide strong leadership. It soon became apparent after the 1974 agreement was negotiated between the union and BCOA, however, that serious concessions had been made to Arnold Miller, particularly in amending the check-off provision and completely rewriting the arbitration procedure.

Both of these changes have cost the coal industry millions of dollars and have resulted in the coal mining state legislatures passing legislation that has caused hardships and has been detrimental to the coal companies and curtailed coal production under the guise of mine safety legislation.

Mr. Camicia's tenure as chairman of BCOA terminated in April of 1976, and E. B. "Ted" Leisenring of Westmoreland Coal Company was named the new chairman of BCOA. Mr. Leisenring's tenure as chairman of BCOA was of short duration. He served from April 1976 through April 1978.

The contract between the UMW and the BCOA expired on December 6, 1977. As a result of the expiration of this agreement, from December 6, 1977 to March 10, 1978, the negotiations for a new contract had come to an impasse. President Arnold Miller found himself in an untenable position. This was the longest strike that the coal industry had ever gone through. The strike covered a period of 111 days. This created a real disruption in the coal industry. As a result, President Carter ordered the Justice Department to go into the federal district court in Washington, D.C. to seek an injunction under the Taft-Hartley Act ordering the miners back to work. The federal court issued the restraining order on March 10, 1978. The union found itself in a very bad situation because it was financially embarassed and faced with having to return to work under the Taft-Hartley injunction.

It is important to note that during the history of the UMWA's existence as a labor union that the Taft-Hartley process has only been invoked by the democratic administrations of President Truman in 1946 and President Carter on March 10, 1978. President Roosevelt seized the mines twice during wartime, but the Taft-Hartley provisions dealing with national emergencies were not at that time in effect.

It became apparent to some of the advisors of President Carter that in some way or other the dispute in the coal industry must be settled because the Democratic party had a presidential election coming up in 1980, and the actions of President Carter would have an impact on the next presidential election. BCOA was determined not to negotiate any agreement with Arnold Miller and would negotiate only on the basis of the requirements of the Taft-Hartley Act. The democratic administration found itself in a bad predicament. In the meantime, President Carter's roving ambassador, Robert Strauss, was requested by the administration to see if he could come up with some type of settlement without going through the process of the Taft-Hartley Act. So, he summoned Nicholas T. Camicia, the chairman of the board and Chief Executive officer for the Pittston Coal Group and Stonie Barker, Jr., president of Island Creek Coal Company of Lexington, Kentucky, to the White House to meet with him. Both of these men were still directors of BCOA, and they were well informed on BCOA's position to let the Taft-Hartley Act take its course in future contract negotiations with Arnold Miller. It was later rumored that Ambassador Strauss made a deal with both Camicia and Barker that if they would break ranks from BCOA and go over to the Mine Workers office at 15 and Eye streets, sit down, and sign a contract with Arnold Miller and Harrison Combs, UMW chief counsel, he would see that the government provided their companies with a large coal order for TVA. Camicia and Barker went to the union's office and signed an agreement with Arnold Miller. After the signing of the agreement, Arnold Miller had a press conference stating that two of the largest companies in BCOA had signed an agreement with the mine workers. At this point, the U.S. Marshal had already served injunction papers throughout the coal industry. After this announcement by Miller, however, the Justice Department was ordered by the Carter Administration to withdraw the injuctions that had been issued by the federal district court. As a result of this action, BCOA signed the agreement. However, many of the companies in BCOA were provoked by the action of the officers of Pittston and Island Creek. This caused a lot of dissatisfaction within the membership of BCOA in what was believed to have been an unethical way by which Camicia and Barker had forced the new agreement upon the other member companies.

Consolidation Coal Company of Pittsburgh, Pennsylvania, the second largest coal company in BCOA, discontinued its membership in BCOA on May 25, 1979. United States Steel, Bethlehem Steel and other steel companies in BCOA also threatened to withdraw their membership because of the manner in which they were forced to sign the 1978 Wage Agreement. For a while, it looked like the beginning of the end of BCOA, so some of the leaders in BCOA prevailed upon the steel companies not to withdraw their membership from BCOA and requested that Consol renew its membership in BCOA. Consol agreed to come back into BCOA if BCOA would allow it to rewrite the bylaws of BCOA. This was the only way that Consol would renew its membership. At a directors' meeting on March 20, 1980, Consol was reinstated to membership in BCOA, and the bylaws were so drafted that the chief executives of the nine largest companies in BCOA, known as the CEO's Committee, would have final say in future contract negotiations with the UMW and would appoint the BCOA negotiators.

The new bylaws creating the CEO's Committee as the one to approve any future contracts between the UMW and BCOA has practically destroyed the effectiveness of BCOA. Many of the smaller companies withdrew their membership, and today BCOA is at its lowest ebb.

The end of Mr. Camicia's employment with the Pittston Coal Group, when he was succeeded as chairman and CEO by Paul Douglas on July 1, 1984, automatically removed him from any future participation in the coal industry. Mr. Douglas was former president and CEO of Freeport-McMoRan, Inc., a natural resources company.

Arnold Miller and his mine workers lost a very close advocate when Camicia left.

Due to his ill health and inability to handle the affairs of the union, Miller made a deal with the International Executive Board and resigned his office as president of the UMW on November 16, 1979.

The union's executive board took official action and named Arnold Miller President Emeritus with a salary of $44,340 plus expenses from 1980 through 1983. During this time, he did no official work for the union. Only a handful of people attended his funeral. He died on July 12, 1985 and is buried in the Montgomery cemetery alongside Route 60 at London, West Virginia.

CHAPTER XXV

The Church Years

After the International Executive Board (IEB) accepted his resignation, the IEB approved Vice-President Sam Church to serve as the acting president of the International Union until the next election of international officers. Church and Miller were both involved in the National Bank of Washington scandal in 1978 wherein $4.5 million was loaned to some Maryland racetrack gamblers and they beat the bank out of this money. This group of gamblers apparently had never intended to pay this money back.

Church officially became the thirteenth president of the UMWA.

"Sam Church was a burly coal miner in his mid-forties from the southwestern part of Virginia, where he had risen to the top position in his local union. There were those who said his success was due more to his fists than to his other qualifications. It was also charged that he had done everything possible to push Arnold Miller out of the UMW presidency so that he, Church, could take over.

"Church suffered from the same handicap as most UMW presidents, lack of education. Like John L. Lewis, he didn't quite finish high school.

"Many persons wondered why Arnold Miller, after the defection of vice-president, Mike Trbovich, accepted Church in his place. Miller may have felt he needed a completely opposite number in his administration.

"In his early months as president, Church made a favorable impression on BCOA officials who thought he approached being a combination of John L. Lewis and the best of Tony Boyle. His propensity for action contrasted favorably with Miller's indecisiveness.

"Belying his early performance, Church used the UMW's Washington bank in a reckless manner to dole out favors to his friends apparently believing, wrongly, that all banks were operated in that fashion. If his derelections with respect to the bank were accurately reported, Federal inspectors should have closed it.

"Church's first serious tests in UMW-BCOA relations came with the renewal of their three-year contract. Negotiations got underway in December 1980 for the March 31, 1981 deadline. On March 23, the negotiators said they had reached agreement, but on March 27, when no contract had been signed, the miners walked out. Shortly thereafter, the UMW's bargaining council approved the contract by a two-to-one vote, but the rank-and-file miners rejected it by a similar vote.

"The miners vote was frustrating to both the BCOA and Sam Church. The BCOA decided to let things cool off before returning to the bargaining table. When negotiations were resumed, Church introduced several new proposals, some of which were summarily rejected by the BCOA. Finally, however, both the UMW bargaining council and the rank-and-file miners approved a contract. The differences between the disapproved and approved contracts were negligible. After seventy-two days on strike, the miners resumed working.

"It was then time to begin thinking about the UWM's 1981 presidential election. Sam Church, aware he had not been elected, naturally desired to have the official approbation of the miners. He announced his candidacy early. His first challenger was his secretary-treasurer, Willard Esselstyn, whose candidacy never got off the ground.

"However, a serious contender from Pennsylvania appeared on the horizon in thirty-three year old Richard Trumka. Trumka did not fit into the customary mold of UMW presidents, particularly with respect to education. He was not only a college graduate, he had a law degree from Villanova University. He worked in the coal mines to get enough money to complete his education. A rumor that he was member of Phi Beta Kappa was incorrect.

"The 1981 election campaign was a bitter one. Church, of course, had to run on his record which Trumka vigorously attacked. Church claimed Trumka's supporters were red-tainted but probably didn't gain many votes thereby. Church was reported to have spent more money than Trumka, but Trumka obviously spent his funds more wisely.

"A significant aspect of the campaign was that UMW membership once over 500,000, was down to 150,000.

"Neither candidate presented any clear view of what lay ahead for the coal industry, or the UMW.

"In the November election, Trumka won by a two-to-one majority."[1]

Thus, Church's term of office was a short one. However, during his term of office he was able to negotiate the National Bituminous Coal Wage Agreement that became effective June 7, 1981. This agreement provided for a wage increase, and also, a pension of $100 per month for widows of deceased members of the UMWA. The widows pension was not in the original settlement which was rejected by the ratification vote of the miners. However, on negotiating further with BCOA, the association agreed to this provision. This contract was probably Church's greatest achievement.

After Church assumed the presidency of the union on November 16, 1979, almost a year later, a couple of articles appeared in the *Washington Post* on October 12, 1980 that were quite detrimental to the officers of the UMWA, and, in particular, Mr. Church. The following articles are quoted from the *Washington Post* and contain more information:

> *National Bank of Washington: A Financial Citadel in Hot Water. Corporate Intrigue, Deception Documented*
>
> Key officials of the National Bank of Washington in recent years have routinely used the city's third largest bank to finance their businesses, real estate ventures and tax shelters—regularly abusing authority and ignoring blatant conflicts of interest, according to thousands of pages of confidential bank records and more than 100 interviews.
>
> Largely cloistered from public scrutiny, a small group of the directors and officers at times bowed to influence peddlers and made unsound loans when it served their own interests and those of their families or friends—often at the expense of the bank's customers and its stockholders, the records show.
>
> The abuses have been tied most often to insider transactions—loans and deals among bank officials using the resources of the bank. Those transactions increased threefold during the last five years, as the United Mine Workers of America, the bank's majority stockholder, forcefully asserted its control in a manner unparalleled in the 30-year history of union ownership.
>
> As a result, the bank has more troubled loans today than ever before, has suffered its worst quarterly profit plunge in nearly a

1. V. B. Harris, *Kanawha's Black Gold and the Miners' Rebellions* (unpublished manuscript), Chapter Ten, Sam Church, UMW President, 1979-1982.

decade and is under investigation by a federal grand jury and the U. S. Securities and Exchange Commission.

NBW's decline is a story of deception and corporate intrigue involving some of the city's most prominent business, civic and political leaders.

During the five-month investigation, *The Washington Post* obtained 3,000 pages of confidential minutes from the bank's executive committee, the small panel of directors that passes on all major loans and bank policies. *The Post* also reviewed thousands of pages of other confidential bank and union documents.

The internal documents and interviews offer a rare glimpse at the inner dealings of a major financial institution and chronicle the impact of a small group of men whose decisions can change the shape of neighborhoods, influence the outcome of political campaigns and pace the growth of business in the nation's capital.

These are some of the findings that the series will report over the next five days:

Federal bank examiners this summer found $53 million in loans with real or potential weaknesses—nearly eight times as much as the bank's annual earnings of about $7 million and more than twice as much as banking experts consider reasonable for a bank of NBW's size. Despite its $860 million in assets, the bank has a net worth of $56 million, leaving a thin margin of reserve in the event of heavy loan losses.

The late Joseph B. Danzansky, NBW's chairman until his death last year, held more than $2 million in insider loans at the bank—some of them personal and some to his business interests—despite a federal law saying that executive officers cannot borrow more than $10,000 from their own institutions.

NBW director and general counsel Ronald G. Nathan, a protègè of Mine Workers President Sam Church, Jr., last year arranged a $50,000 loan for a Washington businessman while acting as the registered agent of the man's troubled business. The loan was a total loss for the bank, but Nathan nonetheless was paid a $10,000 legal fee by the businessman.

Nathan also failed to disclose, as required by federal law, partnerships he held in business ventures that received NBW loans. In one case, 10 days after Nathan signed for his 7 percent partnership in a McPherson Square office building, the project got a $2.1 million renovation loan from the bank.

In 1978, some top officials of the bank agreed to an elaborate procedure to disguise a $50,000 line of credit to former City Council Chairman Sterling Tucker's campaign for mayor by making small loans to individuals who then contributed the money to the campaign. The highly unusual arrangement was used to bypass city election laws that sets the limits on campaign contributions and prohibit banks from making contributions through middle men.

Church, then vice president of the union and a member of NBW's board, told the bank's president to help win approval for a controversial $4.5 million loan to owners of Laurel Raceway in Prince George's County despite credit reports available in the bank stating that the loan might not be repaid and seriously questioning the integrity of the borrowers. The loan was not repaid and the bank lost at least $900,000.

The series of articles starting today documents what federal regulatory officials have feared ever since coalfield titan John L. Lewis secretly bought control of the bank in 1949—namely, that the sometimes volatile union would be unable to resist the occasional temptation to wrest control of the bank from its professional managers, especially when the interests of the union or its officers were at stake.

In the 1950s and 1960s, Lewis used loans from the bank to buy into coal companies secretly, either to prop them up or to force them to accept his union. In the early 1970s, then-UMW boss W. A. (Tony) Boyle used a $179,000 unsecured line of credit to stay out of jail after he was convicted of misusing union funds.

The serious troubles now facing the bank stem not only from a new union leadership forcefully taking control, but also from some overly eager bank managers willing to turn their backs on sound banking principles to curry favor with their union bosses.

The bank officials nurtured close relationships with top union officials, arranging home mortgages for them through savings and loan associations that did business at the bank. The bank paid for memberships in private clubs and picked up the tab for lavish entertainment and recreational trips for union leaders.

The stone facade and turrets of the bank's historic Washington branch at Seventh Street and Pennsylvania Avenue NW no longer symbolize fortress-like strength at NBW.

As a sign of the times, the Washington branch, a national landmark within blocks of the Capitol, was sold to the Argentine Naval Commission last year for $700,000 in a frantic attempt to keep the bank's profits up.[2]

UMW's Power Play: The Union Takes Over the Bank's Board Room

Every Tuesday evening, a sleek black Cadillac limousine leaves the curb outside the downtown headquarters of the National Bank of Washington at 14th and G Streets and weaves its way through Washington's dinnertime traffic to the offices

2. *The Washington Post,* October 12, 1980.

and homes of the dozen men of the bank's executive loan committee.

There are no passengers on the plush seats of the chauffeured car, only a sheaf of papers comprising the minutes of the preceding week's committee meeting.

The minutes are delivered for each director's review before the next morning's meeting. The directors are instructed to return the documents to the bank lest a loose copy find its way elsewhere.

The executive lifestyle comes with the appointment to the board. It means lunch at the exclusive University Club; dinner at the posh and stately Gaslight; soft, rich leather and deep-toned mahogany in the board room. It also means as much as $18,000 per year in extra income for attending the weekly two-hour meetings.

The board room of any bank is in the strictest sense an inner sanctum, where the decisions that shape a community are made in subdued light and rich furnishings. The National Bank of Washington's board room has undergone a radical reshuffling of characters in the last seven years, a virtual game of musical chairs orchestrated by leaders of the United Mine Workers of America, the bank's majority stockholder. UMW leaders have fired six top bank officials—including two chairmen and two presidents—in the past seven years.

The current chapter in the union's history began in December 1972, when Arnold Miller led a reform ticket called "Miners for Democracy" against the corrupt regime of W. A. (Tony) Boyle, the union boss who was convicted of ordering the murder of an earlier rival, Joseph A. Yablonski.

The bank, the union's largest single asset, became an issue in the heated 1972 campaign. True Davis, a former ambassador, aristocrat and socialite who ran the bank for Boyle, had been ridiculed by the reformers for his opulent life style and $142,000 bank salary. Within days of Boyle's defeat, the victorious Miller fired Davis by telegram from Charleston, W. Va.

The first wave of replacements came from the ranks of the Washington business establishment, bringing status and prestige to the bank, along with local business connections.

But with a second group of appointments, beginning in 1978 and 1979, the number of hand-picked men and their friends placed on the board increased at a rate unprecedented in the bank's history.

First, Miller appointed himself to the board along with two other principal union officials, Vice President Sam Church, Jr. and the new secretary-treasurer, Willard A. Esselstyn. For all practical purposes, the union men had virtually autocratic control of the bank, because in any test of power they voted 750,000 shares—a full three-quarters of the bank's outstanding stock.

Ronald G. Nathan, a Miller-Church protègè who had worked in the campaign as fund-raiser and strategist, also was appointed to the board and given the title of general counsel.

Nathan persuaded Miller to appoint to the board two of Nathan's friends, Irving B. Yoskowitz, who had participated with Nathan briefly at Arnold & Porter, and Marvin L. Goldman, president of Washington's K-B Theatres chain, who had contributed $5,000 to Miller's re-election campaign.

At the same time, Church nominated his own close friend and campaign fund-raiser, Willie Runyon, a Baltimore ambulance company owner who came from a family of West Virginia coal miners. In 1979, Runyon nominated his close friend Wilson Lau, a Baltimore insurance executive, followed by Bethesda builder Nathan Landow, and Church added his union administrative assistant, Gary Callen. Corporate lawyer-lobbyist Thomas Hale Boggs, Jr., a Democratic fund-raiser and a friend of Landow, also joined the board.

Arnold Miller resigned the union presidency in the fall of 1979 after a heart attack, and Church took over. Hary Huge, Church's personal attorney, negotiated Miller's resignation. Church subsequently appointed Reba Huge, the lawyer's wife, as an NBW director.

In recent months, Church has intensified the union's ironclad grip on the bank and, effectively, has been acting as chairman, running the bank through a small oversight committee he created, firing officers, appointing new board members and presiding over meetings for the full board of directors. Church has been unavailable for comment. He has denied in a public statement exerting any improper influence over the bank.[3]

Church's troubles began to mount when Richard Trumka defeated the incumbent, Walter Shuba, for international board member from District 4, Uniontown, Pennsylvania. Mr. Trumka is a native of western Pennsylvania, where he grew up in the coalfields. He comes from a family of coal miners and has an undergraduate degree from Penn State University and a law degree from Villanova University. At the time, Trumka, having left his position on the staff of the UMWA law department in Washington, was employed at the Buckeye Coal Company mine in Nemacolin, Pennsylvania. He ran for and was elected International Board Member from District 4. However, there was a question on whether he really met the requirement of five years experience working in or around a coal mine or coal processing plant as required by the International Constitution of the UMWA.

3. *The Washington Post*, October 12, 1980.

The mistake that Sam Church made was not challenging his qualifications when Trumka was elected to the IEB. Trumka was seated on the board and continued to serve as a board member until 1982 when he announced his candidacy to run for the presidency of the UMWA against the incumbent, Sam Church.

Sam Church did not challenge Trumka's membership on the IEB because it was generally known that Church always felt that Trumka was one of his friends.

It developed, however, that Trumka had no use for Church and in 1982 announced his candidacy to seek the presidency of the International Union. In addition to Trumka's announcement to run for the presidency, Willard Esselstyn, secretary-treasurer of the International Union, also announced his candidacy for the presidency. Thus evolved a three-way race between Church, Trumka and Esselstyn for the presidency of the union. Trumka won the election.

CHAPTER XXVI

The Trumka Incumbency

Prior to Trumka's going back to work in the mines, Arnold Miller gave him a letter crediting his employment with the union's legal department which was to be used as partial proof of his mining experience. Of course, this letter couldn't be used because Trumka did not actually have the five years mining experience required under the constitution of the UMWA. There is question whether parttime work in the mines during vacation from school sufficed in fulfilling the five-year requirement.

It is alleged that if he had to prove that he did have the five years experience, he could not even do so today. It is doubtful that his Social Security record would establish five years of employment. It is also commonly known among the anti-Trumka members of the union in District 4 that after Trumka's election, Willard Esselstyn, secretary-treasurer of the union, was asked by the anti-Trumka members of the union to write a check for one dollar drawn on the international treasury and made out to their attorney so that their attorney would have a legal claim against the international union for his services and he could file suit challenging Trumka's qualifications to be president of the international union. This attorney assured the anti-Trumka forces that he would have Trumka disqualified because he did not have the five years experience in the coal mines or coal processing plant.

It is common knowledge that Trumka got Esselstyn to refuse this request, and Trumka promised to take care of Esselstyn as soon as he was installed as president of the union. Mr. Esselstyn has been working for the international union since Trumka's election.

After his installation as president of the union, Trumka got rid of most of Church's supporters.

Although he was from a mining family and possessed a great deal of formal education, Trumka was a very inexperienced person in UMWA and coal industry affairs when he became president of the union. All the experience he had was the short time Arnold Miller employed him to work in the UMWA legal department, his short working experience underground after he left the law department, and during vacations from school. He is most ambitious and envisions himself as a modern-day John L. Lewis. When I was chairman of the Industry Wage Commission, Mr. Trumka was assigned as a staff man to the Mine Workers Wage Commission. Mine Workers Wage Commission members and I had several encounters with Mr. Trumka and had to put him in his place on several occasions.

In May of 1976, I instituted a suit against the International Union in Federal District Court for the District of Columbia for the pension that was due as a result of my eighteen years of employment with the union. This suit was settled out of court between Mr. Trumka and Robert Cadeaux, my attorney. I was awarded several thousand dollars as part of this out-of-court settlement. In September of 1976, I instituted another suit in the Federal District Court demanding that the union reimburse me for payment of dues and assessments from January 1, 1956, the time I terminated my employment with the union, through December 31, 1977, plus ten percent interest and attorney fees.

This case went to trial in the Federal District Court for the District of Columbia, and after the jury was impaneled to hear the case, the attorney for the union, in addressing the jury in his opening statement, stated that he had only one witness in this case and that was Mr. Trumka, who was the attorney for the UMW, and that he would testify that at the time of the settlement of my pension suit with the union, I had assured Mr. Trumka that in the future I would never institute any suits against the union. As a result of this statement by the union's attorney to the jury, the judge declared a mistrial because Mr. Trumka could not appear as an attorney for the union and also as a witness. The judge ruled that in this mistrial, if Mr. Trumka testified, the union would have to engage outside legal representation and they should not use the union's attorney. The case was rescheduled for another date and was tried by a jury in the same Federal District Court and the jury's verdict was handed down in my favor. However, the union did not use Mr. Trumka as a witness nor as an attorney in the second trial. This was later appealed by the union to the Appeals Court for the District of Columbia and the Appeals Court upheld the lower court's decision and I was reimbursed for my dues and assessments including ten percent interest, and the union was required to pay my attorney fees.

Trumka was sworn in as president of the UMW in December of 1982. During the early part of 1983, Trumka began preparing himself, and his supporters, for the International Constitutional Convention that began on December 12, 1983 in Pittsburgh, Pennsylvania. At the convention, several changes were recommended by Trumka. One of the most important changes that took place was that Trumka had the convention abolish the UMW Bargaining Council which had been established by Arnold Miller during his tenure of office, and Trumka replaced it with a Negotiating Committee. This change in the constitution gave Trumka sole authority to be the chief negotiator and to select the persons to negotiate with him. He also recommended that the strike provisions of the constitution be changed, giving him sole authority to call selective strikes when coal companies refused to negotiate an agreement with the union. Trumka was given this widespread authority with no experience in negotiating coal wage agreements and very little knowledge of the workings of the International Union of the UMWA. During the convention, Trumka took advantage of his election by inserting into the constitution a provision giving him broad authority in establishing a Selective Strike Fund by assessing the working miners up to two and one half percent of a miner's daily wages. This fund was to be maintained at a maximum level of $70 million. This was the first time in the history of the UMWA's existence that a provision was written into the constitution to extract money from a working miner's wages for a strike fund.

Trumka successfully negotiated the 1984 Wage Agreement with the BCOA that provided wage increases of 10.25% over forty months and raised the wages of a coal miner from $113 per day to $124 at the end of the contract. This agreement was signed for three and one-half years with an expiration date of February 1, 1988.

The Massey Selective Strike

A. T. Massey Coal Company of Richmond, Virginia withdrew from BCOA about ten months before the expiration of the 1981 contract. Since A. T. Massey refused to sign the 1984 agreement, Trumka elected to institute his selective strike against Massey. Massey's parent companies, Royal Dutch Shell and Morrison Knudson, were among the largest and most financially sound corporations in America. The selective strike theory that Mr. Trumka elected to pursue against A. T. Massey Coal Company has proven to be one of the biggest blunders that the UMW union ever encountered.

A. T. Massey was successful in completely eliminating the UMWA from its operation in Mingo County, West Virginia, and successful in winning several damage suits against the UMW.

It is hard to see how a selective strike can be successful when you do not have the whole industry organized under one union. In 1984, the odds were against Trumka and his union because the UMW represented only 38.6% of the coal produced. A total of 61.4% of the total production of the bituminous coal industry was nonunion, mines affiliated with other unions, and UMWA Western Surface Agreement mines covered by a completely separate agreement.

The only time the UMW were successful in a selective strike was in 1941 under the leadership of John L. Lewis. At that time, however, the union had 99% of the total production under UMWA contract and there was a wage differential between the north and the south of $.40 a day. In other words, the miners south of the Ohio River worked for $.40 less per day than the nothern miners, so Mr. Lewis negotiated a contract with the northern operators for a $1.40 per day wage increase and instituted a selective strike against the southern coal producers. In a few days, the southern coal producers capitulated and signed the national agreement calling for a $1.80 wage increase for the miners, thus eliminating the wage differential between the north and the south.

Mr. Trumka instituted his policy of selective strikes against A. T. Massey and its subsidiaries in the latter part of 1984, and this continued throughout 1988. The agreement between the union and Rocky Hollow Coal Company, a subsidiary of A. T. Massey was signed and became effective October 30, 1988. This agreement would be in force for two and one-half years, or thirty months.

On October 30, 1988, the union agreed to a new agreement covering the employees of Rocky Hollow Coal Company, a subsidiary of A. T. Massey Coal Company, located in Mingo County, West Virginia. The agreement, negotiated by International Vice-President Roberts for the union, was almost a complete change from the BCOA agreement. The Rocky Hollow agreement eliminated the Mine Safety Committee, eliminated payment for most overtime (overtime would only be paid after an employee had worked forty hours per week, eliminated the

April 1 holiday, and many other basic changes. This is a thirty-month agreement which expires April 30, 1991. This was a complete abandonment on the part of the International Union to the miners who worked for A. T. Massey. It is commonly known among the employees that this negotiated contract was agreed upon in order for the company to drop many violations of the federal Hobbs Act that the union was charged with by the company.

The Rocky Hollow Agreement only covers the employees of the Rocky Hollow mine. Only about twenty percent of the employees of Rocky Hollow pay dues. The other employees are replacements for the strikers, who started at the beginning of the selective strike.

The Pittston Selective Strike

At the expiration of the 1984 Wage Agreement on January 31, 1988, there were quite a number of other companies who had also withdrawn from the BCOA and were not covered by the BCOA Agreement that became effective February 1, 1988. There were approximately two hundred large and small companies who refused to sign the new contract. The largest company in this group was the Pittston Company.

Wage negotiations were conducted with these companies from February 1, 1988 through April 5, 1989. The union was able to sign up a few of the smaller companies who had refused to sign the National Bituminous Coal Wage Agreement. However, much of the contract negotiations were conducted with the Pittston Company, and most of these other companies were allowed to continue working without a contract by the union.

The UMW, through its public relations person, Joe Corcoran, announced April 5, 1989, that a selective strike against the Pittston mines in West Virginia and Virginia would begin on the evening shift, April 5, 1989. Eastern Coal Company, a subsidiary of the Pittston Company, and its mines in eastern Kentucky were allowed to continue to work. The strike affected about fourteen hundred employees at Pittston's Virginia operations and about three hundred employees at the Buffalo Creek mine at Lorado, West Virginia and Elkay Mining Company at Lyburn, West Virginiua, or a total of seventeen hundred employees of the Pittston Company.

Picket lines were immediately established at the West Virginia mines on the second shift.

A few days later, Trumka ordered selective strikes against

Gilbert Imported Hardwoods, Lynn Land Company, Belva Coal Company and the Perry-Hylton Coal Company in West Virginia. Selective strikes continued at all of these companies, and the only negotiations were between the Pittston Company and the union.

During the negotiations, the Pittston Company appeared to have been more than fair in presenting to the union its final proposal, in comparison with UMWA agreements with the Rocky Hollow Coal Company, a subsidiary of A. T. Massey and the Benjamin Coal Company, another company with whom the union had finalized a non-BCOA type agreement. A comparison is as follows:

Provisions	Pittston Coal Group Companies	A. T. Massey Companies	Benjamin Coal Company
Union Security	Full Union Security Clause	No Union Security Clause	Full Union Security Clause
Range of Wages	$13.81-$15.87	$11.40-$15.15	$7.80-$8.65
Wage Increase During Contract	$.90/Hour	$.55/Hour	$.65/Hour
Medical Plan	Retirees Covered	Nothing for Retirees	Nothing for Retirees
Pension Plan*	$840/Month	$495/Month	No Pension Plan
Flexible Scheduling	Seven Days Production, Sunday Work Permitted	Seven Days Production, Sunday Work Permitted	Limit of Six Days Production

*20 years service under respective plans, age 62[1]

It was hard to understand why Pittston's offer, as compared to the A. T. Massey Company's and Benjamin Coal Company's was not immediately considered. It is high time that labor in general, and the UMWA in particular, recognize that management has certain inalienable rights that must always be regarded as management's prerogative. The mine workers should confine their interest in collective bargaining solely to reasonable wages and hours and working conditions and not inject itself into those rights of management, which

1. *Charleston Gazette*, May 11, 1989.

must be free to operate if the business is to succeed and continue to provide good employment for the union members.

A large portion of the monies spent by the union during the strike were for activities such as paying for transportation, requiring strikers who receive strike benefits to maintain picket lines where selective strikes against coal companies are being implemented, and to pay for expenses to send mine workers to Greenwich, Connecticut to harass stockholders at the Pittston Company's May 10, 1989 annual meeting to attempt to force the stockholders to adopt three resolutions, which are as follows:

1. A resolution forcing a shareholder vote on a poison pill plan to thwart a hostile takeover;
2. A resolution giving shareholders the right to secret ballot votings, and;
3. A resolution calling for the appointment of outside directors to examine whether Pittston should divest one or more lines of company business.

Its not likely that the union could impose its will when the union and a local only represent a mere 1,721 shares out of the company's 37,800,000. Furthermore, the union had no business trying to intrude into the organizational and commercial functions of the corporation because Pittston hadn't elected to sign the wage agreement proposed by the union, any more than the corporation would want to pry into those functions of the union. To my knowledge, this is the first time in the history of the UMWA that it has proceeded on such a course attempting to destroy a company whose success and profitability are necessary to provide employment opportunity for many of its members.

By June 1, Trumka realized he and his union were in very serious trouble over his selective strike edict of April 5. In order to bolster the position of the union, he dispatched a memorandum to the International Executive Board, detailing the history of his negotiations with the Pittston Company. In this memorandum, he acknowledged this was a life or death struggle for the union. He stated in the memorandum as follows:

> The implications of losing our fight with Pittston. If we lose, Pittston will no longer provide health coverage for retirees, widows and disabled. In all likelihood [*sic*] the burden to provide those benefits would fall to the Funds meaning other employers (Consol, Peabody, etc.) would be forced to increase contributions through the remainder of the contract. They would then rebel, re-

fusing to shoulder other companies' responsibilities. The Funds would go bankrupt and the guaranteed lifetime health care for our fathers, grandfathers and future retirees would be only a memory—over 130,000 pensioners and widows would be affected and could ultimately lose their health care coverage.

The concept of a national coal agreement would fall apart, forcing us to bargain over concessions on a company-by-company, mine-by-mine basis.

All the property and assets of the Union would most probably be seized by the government. Many rank-and-file Pittston miners, District officials, staff and International officers would face civil and criminal penalties.

In short, losing the strike is unthinkable. The consequences threaten the survival of the Union.[2]

On June 11, 1989, a mass meeting was held in Charleston, West Virginia, which Trumka and also some of the union's supporters addressed. Trumka stated "We're tired of having our rights stepped on and our dignity taken; it's time we fought back."[3]

The next day, it is purported that the leadership of the UMW gave out the word to strike coal mines, not only in Virginia, Kentucky and West Virginia, but also in Ohio, Pennsylvania, Indiana, Illinois and Missouri. This was in violation of the national agreement that was in effect in most of these places. These wildcat strikes continued to spread until mines were on strike throughout the coal industry.

Also, Trumka embarked on a most vicious campaign to attempt to destroy Paul Douglas, chairman of the Pittston Company, because Douglas would not heed to Trumka's demand that he personally participate in the coal wage negotiations. Douglas is the current chairman and CEO of Pittston, having succeeded Nicholas T. Camicia in 1984.

Trumka also embarked on a campaign to discredit the Pittston Company and launched a tirade against its officers.

The selective strike against Pittston has been very expensive to the union. Fines were levied on June 2, 1989 against the International Union, UMWA, District 28 and its officers by Judge Donald McGlothlin of Russell County, Virginia. The following is quoted from the court's record:

2. United Mine Workers of America Memorandum from Richard L. Trumka, International President, addressed to All International Executive Board Members in reference to District Conferences dated June 8, 1989.

3. *The Logan Banner*, June 12, 1989.

The Court finds that both of these two associations, the International Union UMWA and the District 28 UMWA, are in contempt with the Court's order, the Court's amended injunction and the order of May 18, 1989, and will command first that the suspended sentences, suspended fines that were previously imposed against those two entities, which were, as I said, suspended, are revoked. That means that the $150,000 of fines that were previously suspended in the case of District 28 shall be paid forthwith—shall be paid in accordance with the schedule that the Court will set forth later. That means that the $250,000 of fines which had previously been suspended in the case of the International Union, that suspension is now revoked and that will have to be paid.

In addition, for the violations that occured on May 8, 11, 12 and 16th, the days before the hearing and order that was orally given by the Court in the last rules to show cause hearing, the Court will impose the following fines on the International and District 28.

For the International $10,000 per day for violations, for a total of $50,000.

For those five days for District 28, $5,000 per day for each day for a total of $25,000 in fines.

For the violations on May 17th, which the Court has found, the International Union is fined $20,000. District 28 is fined $20,000.

For the violations that occurred on May 18th the International Union is fined $50,000. The District 28 is fined $50,000.

For the violations found or committed on the 19th day of May of this year the International Union is fined $150,000, District 28, $75,000.

For the violations that occurred on the 22nd day of May, the International Union is fined $250,000, and the District 28 $100,000.

For the violations found to have occurred on the 25th day of May, the International Union is fined $500,000; District 28 is fined $200,000.

All in all that means that the total fines, including those that the suspension has been revoked on, total $2,020,000 for the International Union, $920,000 for the District.[4]

The International Union appealed these fines to the Virginia Supreme Court of Appeals, and the Supreme Court of Appeals upheld the judge's decision, and the fines were to be paid.

On June 5, 1989, Judge Glen Williams of Federal District Court in Abingdon, Virginia ordered the arrest of three of the

4. Ruth Jones Greiner, C.S.R., R.P.R. (Court Reporter's Transcribed Record of Judge Donald McGlothin's Order of June 2, 1989), pp. 6-8.

international union's top executives. They were strike coordinator, Marty Hudson; the president of District 28, Jackie Stump; and international representative C. A. Phillips. They were ordered to be arrested for violating Judge Williams' injunction on the use of civil disobedience to block coal truck traffic. Judge Williams had the three men placed in jail until such time as they appeared in court and told him that in the future they would obey the law. On June 20, 1989, they appeared before Judge Williams and agreed they would not continue to violate Judge Williams' injunction, and Judge Williams released them from jail.

On June 26, 1989, Judge Williams levied a fine of $240,000 against the union for organizing convoys to slow coal truck traffic.

These court decisions, beginning June 1, 1989, and the fines levied against the union's international representatives, were severe set backs to the union's civil disobedience picketing.

On June 26, 1989, the National Labor Relations Board petitioned Judge Dennis Knapp of the Southern West Virginia Federal Court for a temporary injunction ordering the wildcat miners who were on strike under the jurisdiction of the UMW to return back to work immediately. Judge Knapp issued the temporary restraining order, effective June 26, 1989, ordering the strikers back to work.

In defiance of Judge Knapp's restraining order, President Trumka, on July 1, called for a memorial period of three days at all mines east of the Mississippi River, with the exception of the Westmoreland Coal Company. At the expiration of the three-day memorial period, Trumka extended it for another two days. The miners remained on strike in defiance of Judge Knapp's temporary restraining order.

Trumka was ordered to appear in Judge Knapp's court on August 3, 1989, to show cause why he should not be held in contempt of Judge Knapp's order.

Trumka's so-called selective strike against Pittston was beginning to appear as the largest blunder that the UMW leadership has ever made by influencing its membership to willfully violate the law. The following article is quoted from the *Williamson Daily News* and contains more information:

> (Circuit Judge McGlothlin of Russell County, Virginia) . . . fined the United Mine Workers nearly $4.5 million for "acts of terror" in the coal fields and delivered a blistering critique of union members' violent behavior in the strike against Pittston Coal Group, Inc.

> In Lebanon, Judge Donald McGlothlin, Jr., who earlier fined the UMW nearly $3 million for violating orders limiting strike activities, blasted the union for assaulting and threatening people during the nearly 4-month-long strike.
>
> "There has been a feeling of terror throughout much of this area," the judge said.
>
> He said there has been a "tremendous outpouring" of public money to try to protect the citizens from union violence and a loss of local tax revenues "directly attributable to the violent acts of these defendants."
>
> McGlothlin ordered that $2 million of the fines be paid to the state, $1 million to Dickenson County and almost $1.5 million to Russell County.
>
> He also ordered that all UMW members be restrained from impeding traffic around entrances to Pittston property and from interfering with local traffic.
>
> Many of McGlothlin's fines stemmed from his finding that the union failed to stop "rolling roadblocks," convoys of slow-driving vehicles that have slowed the hauling of Pittston coal.[5]

Judge McGlothlin's fine of $2.9 million on June 2, 1989, was upheld by the Virginia Supreme Court of Appeals.

The union continued to use other methods of harassment to discredit Pittston, particularly the CEO, Paul Douglas as the following quoted newspaper articles from the *Williamson Daily News* will attest:

> On horses and ponies and in horse-drawn carriages, a group of striking United Mine Workers tied up traffic in eastern Pike County (KY) in an off-beat protest against the Pittston Co.
>
> If the company wants to move us back 20 years in the economic standard of living, then we're going to go back 100 years in transportation," said Terry Varney, a spokesman for UMWA Local 5737, which has been striking Pittston for 58 days.
>
> If they want to put us back into the days of horse and carriage, fine. We can get by," Varney said, a reference to Pittston's latest contract offer.
>
> "It's a simple way of showing what they are trying to do to us," said Varney, of the 200 workers striking Eastern Coal Corp., a Pittston subsidiary.
>
> The sounds of horses' hooves striking the pavement and idling car engines mingled as the caravan brought traffic to a crawl on U. S. 119 over Coburn Mountain, a main highway linking Eastern Kentucky with West Virginia. The caravan also slowed coal trucks hauling for the non-union mines in the area.
>
> Varney denied the publicity stunt is a repeat of so-called "rolling roadblocks" used in Virginia's coalfields.

5. *Williamson Daily News*, July 27, 1989.

Last month, a Russell County, Va., judge fined the UMW nearly $4.5 million, in part for the union's failure to stop the convoys of slow-moving vehicles that have slowed Pittston's coal trucks.

Varney said the Kentucky miners were careful to obey the rules of the road when they saddled up and trotted ever so slowly over Coburn Mountain and back again. Kentucky State Police cited only one buggy driver for impeding traffic after he failed to use two turnouts when 75 cars were slowly chugging behind him, Sgt. Don Gill said.

"People were starting to pull out and pass over the double line and the coal trucks were coming the other direction," Gill said, "It was getting too dangerous."

There were no arrests, but Gill said a tire on his police cruiser was flattened by a jackrock after he cited the buggy driver. Jackrocks are clusters of nails welded together and have been used to flatten tires of non-union trucks and police cars.

Gill said horses and animal-drawn carriages are legal on Kentucky highways so long as the carriages are marked with slow moving vehicle signs and the riders use turnouts to let traffic pass.

The caravan braved light showers when it left the intersection of U. S. 119 and Kentucky 199 shortly after 9 a.m. EDT, headed westbound. With one short stop, it arrived about 1½ hours later at an abandoned car wash near Meathouse Hollow, about five miles away.

As the horses and buggies pulled off the highway, a long line of bumper-to-bumper traffic filed past for about five minutes. Drivers appeared weary, but several waved at the miners.

The miners let the horses cool down, and then the caravan headed eastbound at noon over the winding mountain roadway, where traffic is frequently slowed by huge coal trucks lumbering up the grade.

Local 5737 officials kept in constant contact with citizens band radios. More than 100 men were involved in the operation and they were watched closely by nine state police troopers.

The UMWA contends Pittston is trying to break the union by imposing a contract that rescinds job security provisions, cuts health benefits and requires mines to be run all week.[6]

The following is another news account from the *Williamson Daily News:*

T-shirts emblazoned with caustic anti-Pittston Co. cartoons and country music with pointed pro-union lyrics are alternative ways miners are getting word of the Pittston strike out to the general public.

6. *Williamson Daily News,* August 16, 1989.

Wayne LaForce of Norton, Va., was peddling custom designed T-shirts at a recent UMWA picnic in this eastern Pike County community (Hardy, KY). Little was subtle about the shirts, which bore such bold slogans as "Hang'Em High For A Few Dollars More!" and "Corporate Greed, Labor Take Heed."

"One reporter said he spent five days in the coalfields and he learned more in five minutes looking at that T-shirt," said LaForce, 45, a son of a Pittston coal miner.

Each T-shirt tells a pro-union story and is done in a lively political cartoon style. One shirt shows Pittston Co. Chairman Paul Douglas sipping champagne and eating lobster in a jet while below, Virginia state troopers drag striking miners off to jail.

"I've got a personal bias in this," LaForce admitted. "My dad gave 43 years of his life and they won't let him go to the hospital."

The T-shirt also shows Douglas kicking a miner out of the jet with a parachute that says, "Won't open without a contract."

He said many people buy the T-shirts for souvenirs, including some state troopers.[7]

Selective Strikes Against Massey Subsidiaries

On Sunday, August 20, 1989, the union declared selective strikes against Dehue Coal Company, Chafin Coal Company and Big Bear Mining, which were to begin on Monday, August 21, 1989. Dehue Coal Company and Big Bear Mining are subsidiaries of A. T. Massey.

On Saturday, August 19, the pickets at Massey's Rum Creek Coal Sales attempted to stop an end loader that was being used to remove log barriers that had been placed on the road by the pickets, when an altercation occurred as quoted from a newspaper article in the *Sunday Gazette-Mail:*

The vice president of Local 1971 of the United Mine Workers was shot in the shoulder during a Saturday morning confrontation between striking miners and non-union coal company workers in Logan County.

Roy Blankenship, 46, of Amherstdale, was shot during what both contract mine workers and union miners contend was an unprovoked attack by the other side on a road near the Rum Creek Coal Sales mine in Slab Creek Hollow. He was taken to Logan General Hospital and then General Division, CAMC, where he underwent surgery following the 9:30 a.m. shooting at the picket shack in front of the mine entrance.

State police from the Logan detachment said the striking

7. *Williamson Daily News,* August 21, 1989.

Pittston miners and guards both claim the other side fired first. A union miner who was on the picket line said the miners did not have guns.

A release from the A. T. Massey Co., owner of the mine, said six company employees were fired on as they attempted to remove log barriers on the road. The company claimed miners fired 30 shots from three directions, four damaging a coal truck and the endloader's cab, tire and radiator.

"Contractor employees returned gunfire into the hills in self-defense," the company claimed. "An unknown attacker (Blankenship) was reportedly injured by a richocheted shell fragment and taken by ambulance to Logan General Hospital for treatment."

"These vicious attacks threaten the lives of innocent workers only trying to earn an honest living," Rum Creek President Richard Zigmond said. "State police presence and protection is the only way to maintain some law and order in the midst of this guerilla terrorism."

State police, who were patrolling the Slab Fork Road area leading to the mine, said security guards were escorting a front-end loader to the mine when the shooting began. Dozens of rifle and shotgun shots were fired. State Police Sgt. D. L. Belcher of the Logan detachment said troopers were still recovering spent cartridge and shotgun shells late Saturday.

"We just know there were reports of gunshots and each side alleged that each side fired," he said.

The front-end loader also left the road and demolished part of the picket shack, located on a CSX railroad right-of-way near the entrance to the mine.

Miner Henry Farley of Logan was at the shack when he said he heard someone say, "Here comes the 'dozer." He looked up from a bench he was sitting on and saw at least two guards wearing bullet-proof vests and carrying what he described as military-style rifles.

"There were about 12 of us around and in the shack, drinking coffee and eating some doughnuts one of the boys had brought," Farley said. "They was walking in front of the end loader and suddenly they started shooting.

"I ran around to the back of the shack and laid down. I then ran down to where my truck was parked and crouched behind that. They kept shooting and shooting.

"There was this other boy crouched down at the other end of the truck and he suddenly grabbed his arm and yelled 'They shot me in the arm,' and then collapsed.

"We got him in a truck and took him to Logan General Hospital," Farley said. "I have been trying to pick the bullets out of my truck so maybe the police can compare ballistics.

"We don't carry guns. We aren't allowed to. They'd arrest us if we did," Farley said. "This ain't my cup of tea. I'd rather be mining coal."

Miners said more than 100 shots, including tracer ammunition, had been fired into the air from the mine complex late Friday or early Saturday.

State police were trying to determine what sort of bullet struck Blankenship. The bullet shattered into several pieces when it hit him.

No arrests had been made Saturday night and the investigation, patrolling and tension were continuing.[8]

The selective strikes announced by the union on Saturday, August 20 began on Monday, August 21, 1989, and the following article quoted from the *Charleston Gazette* contains more information:

United Mine Workers members began a selective strike Monday at three Southern West Virginia coal company operations, and union protesters wandered the streets of downtown Huntington in an unsuccessful search for one firm's offices.

The union announced the selective strike Sunday against Dehue Coal Co. and Chafin Coal Co. in Logan County and Big Bear Mining in Wyoming County.

A Dehue Coal spokesman said about 40 pickets showed up at a tipple the company operates in Dehue that has been the scene of numerous protests and occasional violent incidents.

"I don't know of any (miners) that tried to report to work," the spokesman said. "They cannot get to work because of trees that are cut across the road."

State police reported little activity throughout the coalfields. "It's super quiet," said Lt. Col. George Young, "almost to the point that it scares us."

More than a dozen union protesters garbed in camouflage attempted to picket Chafin Coal's offices in Huntington, but the targeted building, while named the Chafin Building, has no relations to the coal company, said Steve Rogers, president of the Huntington Chamber of Commerce.

"I looked out my office window this morning and saw people in camouflage attire all of a sudden arrive in front of the Chafin Building," Rogers said. "I found it sort of amusing since the building is not owned by anyone named Chafin nor are there any Chafin-named offices in it."

Rogers said some of the pickets went inside the building, then emerged a few minutes later when their search for the coal company proved fruitless. The group left and showed up not long afterward outside Guaranty National Bank.

8. *Sunday Gazette-Mail*, August 20, 1989.

"Then the news media showed up—Live Eye and everbody else rolled up. I went outside and said "I think you missed them, they went that way' and pointed in the direction they went, he said.

Chafin Coal had maintained an office in west Huntington but Rogers said the company apparently has closed its doors.

At a Sunday news conference, District 17 President Bob Phalen accused the companies targeted by the selective strike of committing unfair labor practices. Phalen charged that the companies have violated contracts which gave UMW members first choice on jobs when a new operator reopens a closed mine.

Union supporters staged protests all last week in Logan County after Con-Serv, Inc., a non-union contractor, reopened the formerly unionized tipple that Dehue Coal operates for Rum Creek Coal Sales, an A. T. Massey subsidiary.

Company president Richard Zigmond said about 160 pickets showed up outside the tipple Monday. Zigmond reported no confrontations.

The union's announcement of the selective strike followed a weekend shooting incident at a Rum Creek deep mine near the embattled tipple that left UMW official Roy Blankenship wounded. Blankenship, 46, of Amherstdale was in stable condition Monday in a Charleston hospital.

State police have made no arrests in the shooting.[9]

Selective Strike Against Pittston—Continued

As reported in the news media with relation to the Pittston strike:

In Virginia, UMW Vice President Cecil Roberts and 18 other union members walked out of court as free men Monday, after a judge dropped contempt charges against them.

Russell County Circuit Court Judge Donald McGlothlin, Jr. said he had seen no legally acceptable evidence that proved the 19 men knew they were violating court orders when they sat down in front of coal trucks at Pittston Coal's Moss No. 3 preparation plant July 25.

But McGlothlin also ordered court officials to serve all 19 with papers detailing the court's prohibition against mass pickets of Pittston operations and against demonstrations designed to slow or halt the Pittston coal traffic, thus making them legally responsible for any future actions.

The 18 other defendants in the case included five members of the union's international executive board, which was preparing to meet in Scott County at the time of the arrests.

Roberts said in earlier interviews he sat in front of the trucks

9. *Charleston Gazette,* August 22, 1989.

to protest the lack of progress in federally mediated talks aimed at ending the 20-week-old strike involving 1,900 Pittston Coal Group miners.

Those talks have been going on sporadically since July 19. Pittston suspended the talks for 10 days Friday because of what it called renewed violence in the Southwest Virginia coalfields.[10]

The harassment of the Pittston Company by the UMWA continued, as can be attested from the following article quoted from the *Charleston Gazette:*

National AFL-CIO President Lane Kirkland, United Mine Workers President Richard Trumka and 15 other labor leaders were arrested Wednesday while demonstrating on behalf of striking Pittston coal miners.

"I do not feel the need to be arrested at any time, but I do feel the need to show my support for my brothers and sisters," Kirkland said while being processed at the city jail. "If that's a misdemeanor, then so it is. . . . If it's a misdemeanor, then I have to plead guilty."

More than 2,500 arrests have been recorded since the miners walked out April 5, with most stemming from the mine union's unprecedented use of civil disobedience sit-ins.

Pittston, which argues it needs a contract with more flexibility to compete with overseas markets, broke off talks last week to protest what it called an increase in union violence. The dispute is more over benefits and scheduling than wages.

The labor leaders addressed a Wednesday morning rally at the state Capitol in Richmond, and after the entourage left, strikers set up tents and vowed to stay until Virginia Gov. Gerald Baliles met with them.

Baliles, who is on vacation in Virginia Beach, authorized spending $1 million on state troopers in the coalfields and has not set foot in the area since the strike began.

Union members argue the use of troopers to escort coal trucks belies his official neutrality stance. Police gathered to clear the Capitol square, but were not expected to make a move before the grounds closed at 11 p.m.

Pittston last year mined 13.6 million tons of coal in Virginia, West Virginia and eastern Kentucky, making it the nation's 15th largest coal producer. The coal was produced from large, unionized underground mines operated by the company's Pittston Coal Group subsidiary, and by the non-union surface mines of Pyxis.

Last year, Pittston sold 7.8 million tons of its coal to the steel industry. Most of that, plus some additional coal the com-

10. *Charleston Gazette,* August 22, 1989.

pany bought from other producers, went to steel companies in Japan, Korea and Canada.

Such high quality coal brings premium prices, but it's also expensive to produce, because it most often comes from deep seams that are harder to mine. It also must compete with coal from Australia, Colombia and other countries, limiting profit margins, said Charles Haywood, a University of Kentucky economist and a Pittston director.

The steel industry has been relatively strong the past year or two, he said, and that has improved the demand, pricing and profits for Pittston's metallurgical coal, which is used in steel production.

But an economic slowdown would change that. And company officials believe Pittston's long-term growth in the coal industry must come from sales to utilities.

Also, ways of controlling and predicting costs more effectively are critical if Pittston is to grow in the highly competitive world markets of the 1990s, Haywood said.

Company proposals for providing a level of job security for union miners and for redesigning their health plans would achieve that, Haywood said.

Haywood said he and his fellow directors are solidly behind company chairman Paul W. Douglas in seeking those goals.

Most stockholders, so far, also seem to be supporting Pittston's tough line with the UMW. The price of Pittston's common stock recently has ranged around $20 a share, down only about $2 from its 52-week high.

That indicates shareholders still take a long-term view, and will accept lower profits for a while because they believe the future will be brighter, Villagran said.

"But at (some) point that is going to swing," Villagran said, with shareholders and perhaps Pittston customers losing faith.

That hasn't happened yet, though, Villagran said, and he sees no end in sight for the Pittston-UMW dispute.

A Baliles spokeswoman said the Democrat saw no need to meet with them because he met with Trumka last week.

A few hours after the rally, in mountainous coalfields 250 miles from Richmond, Kirkland, Trumka, Virginia AFL-CIO President David Laws, 14 top labor leaders and five other supporters sat down to block the double doors of the Russell County Courthouse.

After refusing to disperse, they were arrested and charged with a misdemeanor—obstructing the free passage of others. After posting a $500 bond, the men headed to the strikers' weekly Wednesday night rally.

All face a year in jail, a $1,000 fine, or both when they come to trial Dec. 5.

Kirkland conceded the arrests were largely symbolic but said the group wanted to show "solidarity with our brothers and sisters in a great union, the United Mine Workers."

Others arrested include Gerald McEntee, president of the American Federation of State, County and Municipal Employees, and the presidents of the transport workers union, the electrical workers union, the airline pilots association, the union representing aerospace workers and machinists, the department store workers union and the head of the international union of service employees.

When the group was led away to jail, several hundred strikers—the majority wearing the camouflage fatigues they've adopted as a strike uniform—stood and cheered or joined in a horn-blasting motorcade that jammed downtown streets.

The Pittston corporate umbrella covers Burlington Air Freight and the Brinks home security and armored car companies, but the coal group provided roughly half of the parent company's profit in 1988.

Coal tax records suggest coal production has been cut in half by the strike despite the use of replacement workers. Some 1,900 miners are on strike in Pittston operations in Virginia, West Virginia and Kentucky.[11]

The following quoted newspaper article from the *Williamson Daily News* tells what the outcome of the strike against Pittston means to coal's future:

The path the Pittston Co. follows in settling its five-month-old labor dispute with the United Mine Workers will affect the coal industry for years to come, according to industry officials.

That's because the company wants to stop contributions to a health plan for 119,000 union retirees and families.

If that happens, other companies are expected to follow, and that could soon eliminate the Medicare supplementary insurance for the group. Its loss would irreparably damage the union and its current leadership since it has been a leading benefit for the union for more than 40 years.

But the strike also could seriously damage Pittston. And at least one analyst, Rafael Villagran of Shearson, Lehman, Hutton, Inc., believes a failure by Pittston to win the cost reductions and increased flexibility it wants could eventually take the company out of the coal business.

More immediately, such failures would slow Pittston's expansion in the production of coal for U. S. electric generating companies, Villagran said.

To do that, Pittston wants to expand the profitable surface

11. *Charleston Gazette*, August 24, 1989.

mines it has begun operating within the last five years through its non-union Pyxis Resources, Inc. subsidiary, Villagran said.

Managers want to win a contract with the UMW that will allow them to keep expanding and acquiring operations of that type without union involvement.

Pittston spokesman Bill Byrne said the company's Pyxis operations will remain non-union, because they operate well that way. He also said the company is considering expansions into other kinds of natural resources, such as copper or other metals and minerals.

Coal, however, should remain in Pittston's future, no matter how the UMW dispute is ultimately resolved, he said.

Company officials believe that more flexibility in its contract will allow its UMW mines to become more efficient, Byrne said.

The company also might take on more employees, adding to the nearly 1,700 miners employed in Pittston mines before the strike, Byrne said. He said the mines won't grow without such changes.[12]

The importance to Pittston, from a cost-competitive standpoint in resisting the union demands that it contribute to the industry-wide pension and health and welfare funds and subscribe to the BCOA Agreement finds its basis in its own economic basis. To put this into perspective, the following quotation from *Forbes Magazine* appears pertinent:

Back when he was steering $2.2 billion (sales) Pittston Co. to record profits, they called Nicholas T. Camicia "Mr. Coal." That was seven years ago, when Pittston, the nation's largest independent producer, was riding high. Profits jumped from $25 million in 1973 to $200 million in 1975.

Now, many of Camicia's competitors are posting surprisingly good profits—but not Pittston. The company earned only $20 million for the first half of this year on sales of $1.3 billion. That's down 56% from 1980, the last comparable period. Pittston wasn't able to cover its dividend last month. It was the first dividend cut in company history.

The problem is obvious: Camicia, age 66 and the son of a coal miner, who rose through the ranks at Island Creek Coal to become Pittston's chairman, profoundly midjudged his business.

He relied too heavily on the metallurgical coal market, which is used in the depressed steel industry, and gambled on the prospects of selling environmentally acceptable coal to utilities which today is still a hazy dream.

Not all of Pittston's problems are Camicia's creation, of course. His predecessor's diversifications are floundering. The

12. *Williamson Daily News*, August 28, 1989.

oil distribution group produces half of corporate revenues, but margins are barely 2% and falling. Then there's Brink's, the security company, another business in which profits are under pressure. Finally, Camicia's attempt to move into non-capital-intensive activities—through the purchase of Burlington Northern Air Freight five months ago—now looks ill-timed.

Pittston's plight goes back to the late Seventies. As its competitors were shutting down their met coal production and digging steam coal mines, Camicia was plowing $225 million into new facilities that largely serve the met market. Today there is domestic over-capacity of some 30% in met coal and, given the state of the U. S. steel industry, the domestic market might never return.

What's more, the export market is soft. Pittston's major customer, Japan—which normally takes 25% of its U. S. coal imports from Pittston—is heavily overcommitted. The company recently shut down eight mines, two permanently. "It's a very bleak situation," Camicia admits, predicting growth in the met coal market of less than 3% a year.

Some of the trouble stems from "compliance coal," a low-sulfur, top-quality product that—Pittston hoped—would be forced on utilities by the EPA. That didn't happen, instead, regulators required new utilities to install scrubbers. While Camicia predicted two years ago that the compliance coal market would reach 25 million tons annually, he now pegs it at 10 million tons.

Back when he was excited about compliance coal, however, Camicia suggested that some of Pittston's assets might be sold. There was talk, too, of a joint venture with an overseas company. That started takeover rumors—and the company's stock price jumped, moving from 18 to 31 in 1980. Camicia pulled in $741,000 by exercising a stock option when shares were close to that year's high. Pittston shares recently traded at 13 3/4.

Camicia's latest controversial move was his May acquisition of Burlington Northern Air Freight. Pittston paid $177 million in cash—borrowed cash—which most analysts think was too much. "How can you pay $150 million more than the asset value of a company?" wonders one industry analyst. This year, and quite possibly for the next several years, BNAFI's earnings will fail to cover Pittston's interest expense and the amortization of goodwill.

At best, Pittston seems to have bought BNAFI near the peak of the market. Earnings, which averaged 21% growth since 1979, are flat. What's more, the fundamental goal of the acquisition—to move into a non-capital-intensive business—is now in question. BNAFI, which has always refused to own its fleet of planes, may not be able to continue that arrangement, as the freight forwarding business grows more competitive. Camicia dismisses that possibility, but admits that if Pittston had to buy planes, "I guess we would do it."

There are credibility problems at Pittston, too. Critics feel the company has been less than straightforward with the investment community. Last quarter, for example, Pittston was hit with $7 million in charges connected to mine closings. Not only was this information not included in the quarterly press release, but analysts were kept in the dark until several weeks after the fact.

One explanation for the confusion could be the fact that five top executives have left Pittston over the past two years. Among the departees were two executive vice presidents, then the company's second-highest paid officers. Did all the turnover disrupt management? Camicia doesn't think so. "The only turmoil was in the mind of the guy who got fired," he says.

That attitude reflects Camicia's self-confidence—something that Pittston's directors have never seriously challenged. Three years ago they handed Camicia a seven-year contract, which will expire just before he turns 70. The board was eager to establish a plan of succession, and last year Camicia tapped Hendrik Hartong, Jr., then president of Brink's, to become president and chief operating officer.

These days the two are reportedly feuding, even though they deny any struggle. Industry observers, meanwhile, aren't terribly keen on seeing Hartong as Pittston's next chairman. He had no coal experience when he came to Pittston in 1977, and until last year he had nothing to do with Pittston's coal group.

A fresh approach, of course, may be just what Pittston needs. After all, Camicia—the former "Mr. Coal"—has made some major blunders in his time.[13]

Wage negotiations between Pittston and the UMW continued under the auspices of the Federal Mediation Service in Duffield, Virginia. However, Mr. Trumka continuously attacked the integrity of Paul Douglas, chairman of the Pittston Company, and charged his company with attempting to destroy the UMWA. This propaganda, promoted by Mr. Trumka, was not very effective because he alleged that if the UMW lost this fight with Pittston, other industries would attempt to destroy unions in those industries. There is nothing in the federal labor laws that forces any employer to sign a contract with a union unless it desires to do so. The employer has the same rights as a labor organization, and if we truly believe in a democracy and human rights, we certainly cannot condone the

13. *Forbes Magazine*, October 11, 1982, pp. 41-42.

destructive practice of the UMWA that plagued the Pittston strike since April 5, 1989.

Meanwhile, Trumka continued his harassment of the Pittston Company by having striking employees of Eastern Airlines and members of the United Automobile Workers union travel to the Pittston operation in Virginia in large caravans to agitate the pickets against the Pittston Company.

This caravan arrived in Virginia on Saturday, September 9, 1989, and along with this caravan were 240 religious leaders. Surprisingly, the religious leaders who came to Virginia to exemplify their solidarity with the UMW; who profess to be religious leaders and know what is right and wrong, refused to comment about the mine workers appearing in camouflage uniforms, and destroying property. Hundreds of them were arrested for violating the laws of the state of Virginia and for spreading literally hundreds of jackrocks on the highway to destroy the tires of the coal trucks. They did not blame any of the union's officers, including Mr. Trumka, for advocating that the miners picketing Pittston violated the laws and broke windshields on the trucks and cars of the officials of the Pittston Company. These religious leaders remain silent not only about those arrested and prosecuted for violation of the law but also the fines that were imposed against district and international officers and organizations for these violations.

In some cases involving the district and international (including their officers) in which heavy fines were assessed against them, these fines were appealed to the Virginia State Supreme Court of Appeals, and these appeals were denied by the Supreme Court of the state of Virginia. These activities, on the part of the union, did not improve their image, and in fact, their image among the general public reached a very low state.

The UMWA suffered a large blow from the courts in Russell County, Virginia, and also the Federal District Court in Big Stone Gap, Virginia wherein ninety-eight members of the UMW, and one clergyman took over and staged a sit-in at the Pittston Moss No. 3 preparation plant and chained themselves in the plant. As a result of this takeover, on September 18, 1989, the attorneys for the National Labor Relations Board petitioned Federal Judge Glen M. Williams to take formal action against the officers of the union and their members, and particularly those who were involved in the sit-in at the preparation plant. On September 20, 1989, Federal Judge Glen M. Williams issued his order as requested by the National Labor Relations Board. Judge Williams' order is as follows:

This cause is again before the Court upon a Motion filed by petitioner and an Order entered on the 18th day of September, 1989, summoning the International Union, United Mine Workers of America (International), District 28, UMWA (District 28), and the following named individuals:

> Cecil A. Roberts, International Representative,
> United Mine Workers of America,
>
> John Cox, International Representative,
> United Mine Workers of America,
>
> Rick Blaylock, International Representative,
> United Mine Workers of America,
>
> Eddie Burke, Regional Director Organizing,
> United Mine Workers of America,
>
> Jackie Stump, President, District 28,
> United Mine Workers of America

herein collectively referred to as the Individual Contemnors to show cause why they should not be held in civil contempt of this Court's June 7, 1989 temporary injunctive decree issued pursuant to Section 10(j) of the National Labor Relations Act ("the Act") (29 U. S. C. Section 160(j)).

A hearing was conducted pursuant to said Order on September 20, 1989, at Abingdon, Virginia. Upon the testimony, exhibits, and arguments of counsel I find that on September 17, 1989, in the late afternoon, the International, by and through its agents and representatives, including Cecil A. Roberts, International Representative, and John Cox, International Representative, and, acting in concert with the International and its agents Roberts Cox were District 28, its President, Jackie Stump, Strike Coordinator Eddie Burke and International Representative Rick Blaylock, all of whom gathered with others en mass at the Employer's premises at Moss No. 3 Preparation Plant, Russell County, Virginia. A group of approximately 100 Union members, including Individual Contemnors Eddie Burke and Rick Blaylock occupied the Employer's premises at Moss No. 3 at that time and refused to relinquish control of the premises back to the Employer since that time. In addition, as many as 1,000 Union members and persons acting in concert with them congregated in front of the entrances to Moss No. 3. As a result of this physical occupation and mass demonstration, employees of the Employer have been unable to enter the premises and perform their jobs at Moss No. 3. The Union's occupation of Moss No. 3 is ongoing. The Union, through the actions of its agents and representatives and through prepared statements, has taken full responsibility for these actions.

On June 7, 1989, this Court issued a Section 10(j) injunction ordering the International, its officers, agents, representatives, servants, employees, attorneys, and all members and persons acting in concert or participation with them, to cease and desist from, inter alia:

> (a) mass picketing; physically blocking the ingress to and egress from the Charging Party's ("the Employer's") job sites in Wise, Russell, and Dickenson Counties, inflicting and attempting to inflict damage to the property of the (Employer) and its employees, including specifically motor vehicles owned, operated, driven by or carrying employees; following employees, inflicting and attempting to inflict damage and/or injury to the property and person of employees, or in any other manner obstructing, hindering, preventing or stopping or attempting to obstruct, hinder, prevent or stop any employee of the (Employer) or persons doing business with the (Employer) from going to or from their place of employment or from working:
>
> (b) in any other manner restraining or coercing employees of the (Employer) or persons doing business with the (Employer) in the exercise of their rights (under the Act).

On June 26, 1989 and July 5, 1989 this Court found that the International and its agents John Cox and Cecil A. Roberts had engaged in certain conduct violative of the Court's June 7, 1989 temporary injunction; the Court had adjudicated the International and these individuals in contempt, imposed certain monetary sanctions and established a schedule for future monetary sanctions for future conduct in violation of the Court's injunctive order.

Accordingly, the Court is of the opinion and hereby finds that the International, District 28, and the Individual Contemnors are in violation of this Court's Orders.

WHEREFORE, in order to purge themselves of contempt, the International, District 28, and the Individual Contemnors are hereby ordered to take the following actions:

1. The International and District 28 shall direct their members to vacate the Employer's facility at Moss No. 3 Preparation Plant and to cease congregating at the entrances thereto on or before 7:00 p.m. September 20, 1989. The International and District 28 shall further inform their members that failure to do so will result in imposition of internal union disciplinary measures and sanctions against them and the denial of any future strike benefits and, upon such failure, take those actions. The International shall so direct its members through its President, Richard L. Trumka. The International shall also read aloud to all its members at Moss No. 3 the injunctive order issued by this Court on June 7, 1989 and inform its members that neither it nor District 28 will provide bail bond nor legal expenses for

any members who are arrested for failure to vacate Moss No. 3 or for any other future violations of the Court's injunction.

2. The International shall provide compensatory damages to the Employer for all losses stemming from its failure to comply with this Court's order by its occupation of Moss No. 3, provided, however, that the imposition of such damages will be suspended if the premises at Moss No. 3 are vacated on or before 7 p.m. September 20, 1989.

3. The International shall pay fines for each day the occupation continues in accordance with the fine schedule established by this Court on July 5, 1989. Moreover, that fine schedule will be prospectively increased by $100,000 to $600,000 for every day of any future contumacy by the International, its officers, agents, or persons acting in concert therewith.

If the International, District 28, and named Individual Contemnors fail to comply on or before 7 p.m. September 20, 1989 with any or all of the above affirmative relief, any of the named Individual Contemnors shall be immediately subject to a writ of body attachment until they have complied with the terms of this purgation order and the terms of this Court's original June 7th injunction.

4. District 28, UMWA, shall pay fines in the amount of $25,000 per day for each day of continuation of contumanious conduct.

5. Cecil A. Roberts and John Cox shall pay fines of $50,000 a day for September 17, 18 and 19, 1989, the days they have engaged in contumanious conduct under the extant fine schedule set out in the Order of this Court dated July 5, 1989, and, prospectively, said fine is increased to $60,000 a day if either of them engage in further contumanious conduct.

6. Prospective fines in the amount of $10,000 per day for each day of further contumacious conduct shall be paid by Jackie Stump, Rick Blaylock, Eddie Burke and any International, District 28, or Local Union officer or agent.

7. All fines assessed herein, and prospective fines, shall be paid to the National Labor Relations Board.

8. Neither the International nor any other United Mine Worker entity shall reimburse any or all of the fines imposed upon any current or future contemnor.

9. The International shall reimburse the National Labor Relations Board for the costs it incurred in investigating and prosecuting this contempt proceeding.

10. It is further ordered that the International and District 28

(i) Give to each of its members engaged in picketing in Wise, Dickenson and Russell Counties, Virginia as well as all of their officers, representatives, agents, servants, employees and all persons acting in concert or participation with them, including

Local Unions, a clear written directive to specifically refrain from engaging in any further conduct in violation of this Court's Order of June 7, 1989.

(ii) Within ten (10) days of this Order, file a sworn affidavit with the Regional Director of Region Eleven describing with specificity the steps taken to assure compliance with the terms of this Order by their members, officers, representatives, agents, servants, employees and all persons acting in concert or participation with them.

It is further ORDERED that, pursuant to the nationwide service afforded the U. S. Marshal's office by Section 11(5) of the National Labor Relations Act (29 U. S. C. 161 (5)), the United States Marshal's Office shall serve this Order and notice upon the abovementioned parties forthwith and the Clerk shall deliver a copy of this Order to counsel of record.[14]

It was inconceivable that Mr. Trumka, who is a lawyer by profession, would encourage these pickets, who are members of his union, to continue to violate the federal and state laws of Virginia and ignore the decisions of the courts, particularly this order by Judge Williams.

In a speech given to a rally on October 8, 1989, in Logan, West Virginia, Trumka made the following statement in reference to this takeover. "I extend a special thanks for the non-violent union takeover of the prep plant in Virginia where 17 members of UMWA Locals 1971, 5815 and 8454 were stockholders inside the Pittston Moss No. 3 Preparation Plant."[15] It seemed that these enormous penalties for violation of the law had no effect on the manner in which Mr. Trumka addressed the members of his union who were carrying out his instructions in picketing Pittston Company operations.

The fines by Judge McGlothlin continued to mount. The following article quoted from the *Bristol Herald Courier/Virginia Tennessean* contains more information:

> The amount of fines imposed on the United Mine Workers union by Russell County Circuit Judge Donald McGlothlin Jr. has now surpassed $30 million.
>
> The judge has assessed the union another $6.9 million in fines after finding the UMW "guilty of 71 violations (of a court order) involving violence" between early July and mid-August.
>
> During Wednesday's hearing before McGlothlin, a UMW

14. Order of Judge Glen M. Williams in the U. S. District Court for the Western District of Virginia, Abingdon Division, Civil Action #89-0090-A, September 20, 1989.

15. *The Logan Banner*, October 9, 1989.

member was charged with throwing a 2-foot-tall ashtray can through a plate-glass window into a room full of Pittston witnesses, two of whom had to be taken to a Lebanon hospital for treatment of cuts from broken glass.

Franklin D. Cantrell, 53, a striking Pittston employee from Jolo, W. Va., was charged with a felony count of maiming and misdemeanor counts of destroying public property and trying to intimidate witnesses. McGlothlin added a contempt charge. Cantrell was released about 7 p.m. after making bond.

The witnesses were waiting to testify on charges made by Pittston against the union regarding violent acts that violate McGlothlin's injunction regulating striking activities.

McGlothlin made his ruling Thursday evening as three weeks of contempt of court hearings concluded.

The judge ruled that $2.5 million of the penalty should go to the Commonwealth of Virginia, $1.8 million should go to Russell County, $1.2 million should go to Dickenson County and $1.4 million should go to Clinchfield Coal Co.

The union went on strike against Clinchfield and other Pittston Coal Group companies April 5.

The union has filed a motion for McGlothlin to reconsider the $3.4 million fine, which the judge made payable to the company, in a Sept. 21 order. A similar action is expected on this round of fines.

The judge said the $1.4 million earmarked for Clinchfield should be paid by 5 p.m. on Friday, Oct. 13.

In his closing, company lawyer Stephen Hodges said making fines payable to the company is the method of enforcement "most likely to bring the union in compliance with the court's order."

McGlothlin has assessed $23.8 million worth of fines against the UMW during the past three weeks.

The judge said copies of the court order must be sent to UMW President Richard Trumka, vice president Cecil Roberts and secretary-treasurer John Banovic, as well as all members of the union's International Executive Board. The order "shall be read at all meetings" of this group of union officers, McGlothlin said.

Despite public statements by Trumka and other UMW leaders advocating non-violence, Hodges said the union is advocating one policy in the press and another on the coal haul roads of Russell and Dickenson Counties.

Hodges indicated large groups of UMW members are taking part in rock-throwing. He said such activity is being organized at sites such as a ballfield in the Cleveland area.

The company attorney also noted testimony by a Pittston supervisor who said he'd seen several UMW "strike captains" among groups while rock-throwing was taking place. While these UMW leaders aren't throwing rocks themselves, their presence indicates the event was sanctioned by the union, Hodges said.

Union lawyer Jim Haviland asked McGlothlin to take into consideration "the massiveness of misery that's resulted from this strike," as well as the alleged unfair labor practices with which the companyh has been charged.[16]

Mr. Trumka continued to seek the support of politicians to justify his actions against the Pittston Company. He engaged Senator Rockefeller of West Virginia in attempting to have legislation enacted to amend the law to compel the trustees who administer the Health and Welfare Fund for the union to force Pittston to pay into the health and benefits programs.

The following article quoted from the *Charleston Gazette* contains more information:

Pittston Coal Co. President Paul Douglas said Monday that a bill introduced by Sen. Jay Rockefeller, D-W. Va., to keep the United Mine Workers' struggling health benefit trust funds afloat amounts to "anti-competitive collusion."

West Virginia's junior senator responded with some tough talk of his own. "That's about what I'd expect from Paul Douglas," Rockefeller said. "He's trying to make other people pick up health benefits that he ought to be paying."

Rockefeller also made no bones about the secondary intent of the bill—to bring an end to the bitter 6-month-old labor dispute between Pittston and the UMW.

"If the bill passes, Pittston is not only going to be substantially poorer, they're going to have substantially more reason to settle," Rockefeller said. "We're trying to put a lock on the bear trap. They've jerked our people around long enough."

A provision of the Coal Industry Health Benefit Stabilization Act of 1989, introduced Friday, would require Pittston and other companies that paid into the UMW's pension and benefit trusts prior to Jan. 1, 1988, to continue that commitment. The National Bituminous Coal Wage Agreement signed by the union and the Bituminous Coal Operations Association became effective on that date.

The UMW has been on strike against Pittson over those same benefits. After Pittston withdrew from the BCOA and refused to sign the Coal Wage Agreement, the company stopped making payments to the trusts.

Douglas accused the UMW and large coal companies of trying to bankrupt smaller mining operations. He called the bill a "private bailout" arranged for their convenience.

16. *Bristol Herald Courier/Virginia-Tennessean*, October 6, 1989.

Earlier in the day, Pittston spokesman William Byrne said the company consulted legal authorities who believe the bill "is clearly unconstitutional."

Byrne also said the proposed legislation, should it become law, would have a "very, very detrimental effect" on negotiations between Pittston and the UMW "for the obvious reason that it would take one element out of the negotiations."

Rockefeller said that's precisely the point. He said the UMW and the BCOA approached him about introducing the legislation at the West Virginia Chamber of Commerce's annual meeting at The Greenbrier in early September. He said the discussions included UMW President Richard Trumka, BCOA President Joseph Brennan and BCOA Chairman Robert Quenon.

The senator said it was at his suggestion that the proposal include a provision mandating retroactive payments into the funds, a provision aimed squarely at Pittston.

"Supposing Paul Douglas gets away with this, what does that say for the future health of the trust funds in terms of signatory companies paying into it?" Rockefeller said.

"Who would be the most hurt if Pittston gets away with this sham that they're trying to pull? The most vulnerable segment of the population. The stakes are extremely high on this."

The bill also carried the signatures of Sen. Robert C. Byrd, D-W. Va., as well as Pennsylvania Sens. John Heinz and Arlen Specter, both Republicans. Rockefeller said West Virginia's congressional delegation is solidly behind the proposal.

"We intend to go to war on this thing," he said.

In remarks he prepared for a Senate floor speech Monday, Rockefeller said the proposed legislation would allow the transfer of money from the UMW's overfunded 1950 pension trust to its 1950 and 1974 health benefits trusts.

The two trusts provide health benefits to 125,000 beneficiaries, including about 32,000 UMW pensioners in West Virginia.

The 1950 fund has a $54 million deficit and the other is expected to be in deficit within weeks," Rockefeller said. "As a consequence, payments to doctors, hospitals and pharmacies are overdue by as much as 60 days."

The UMW and BCOA would jointly decide the amount of money to be transferred between the funds. "This is a critical point, because it means that union representatives will have coequal participation and veto power, the senator said.

Trumka and Brennan both expressed their support of the bill in a press release issued by Rockefeller's office.

"There is an urgent need to stabilize the financial condition of the retiree health care programs in the coal industry," Rockefeller said in a press release. "Without this legislation, it is likely that the retiree health care program will fall further into the red, jeopardizing health benefits for thousands of coal miners.

"By stabilizing the retirees health care system, we will help maintain the competitiveness of the U. S. coal industry in the world marketplace."

Pittston said the legislation would destroy the competitiveness of the domestic coal export industry. The company said the bill "sets a precedent for revolutionizing the collective bargaining process by permitting a select group of large, private companies to impose their bargaining goals on an entire industry."

The bill also provides statutory enforcement of UMW-BCOA contract provisions requiring payments for retiree health benefits. Rockefeller said he expects action on the bill before Thanksgiving.[17]

After the Senate Finance Committee approved the Rockefeller bill and in order for this bill to become law, Senator Rockefeller attached the bill to the reconciliation bill that was before Congress as part of the federal budget to be passed. To Rockefeller's surprise, during the week of October 9, 1989 the U. S. Senate killed the reconciliation bill, and therefore, Rockefeller's Coal Industry Health Benefit Stabilization Act of 1989 was defeated. This was a victory for the Pittston Company and the other smaller coal companies that opposed this legislation throughout the industry.

In October, the federal government stepped into the picture. The following article quoted from the *Williamson Daily News* contains more information:

Labor Secretary Elizabeth Dole did her part to get the heads of The Pittston Co. and the United Mine Workers to meet for the first time since the rancorous strike began six months ago.

Mrs. Dole said she hopes a meeting she arranged this afternoon between Pittston Chairman Paul Douglas and UMW President Richard Trumka will mark the beginning of the end of the labor dispute.

"I'm not a mediator. I'm not here to say I have the answer to this complex problem, but I do feel now it is time to see if we can facilitate the bargaining process," she said. "Ultimately, it has to be resolved between the private parties."[18]

On October 25, Labor Secretary Dole announced further efforts to end the Pittston strike. The following article quoted from the *Wall Street Journal* contains more information:

W. J. Usery, Jr., labor secretary during the Ford administration, was named to mediate talks to settle the six-month strike by the UMW.

17. *Charleston Gazette*, October 3, 1989.
18. *Williamson Daily News*, October 14, 1989.

Previous talks between Pittston, of Greenwich, Conn., and the union have been sporadic and unsuccessful.

The union called the strike in April after Pittston refused to sign the UMW's national labor pact. Pittston seeks changes in health and pension benefits, among other things. No schedule for formal talks was set, but meetings are expected to begin soon.[19]

Since an impasse was declared in the negotiations between the union and the Pittston Company on June 7, 1989, under the National Labor Relations Act this gave the Pittston Company authority to implement its own agreement with its employees, and on June 10, 1989 Pittston implemented its own agreement providing for higher wage rates and pensions than are paid in the present UMW agreement and established its own pension, health and benefits program. This agreement covered more than 750 employees who were hired as replacement workers during this strike.

On October 30, 1989, Judge Donald McGlothlin of Russell County, Virginia Circuit Court, as reported in the *Williamson Daily News:*

> . . . postponed additional fines against the union to allow contract talks to move forward with Pittston Coal Group, Inc.
>
> The judge said he wanted to avoid any action that could impede progress in contract talks mediated by former labor secretary Bill Usery.
>
> McGlothlin said the union violated court-ordered strike restrictions 84 times in August and September, and that nearly half of the acts were violent.
>
> The most blatant act of contempt of court the judge cited occurred last month when 98 union members invaded Pittston's main preparation plant and occupied the building for four days.
>
> McGlothlin previously issued fines totaling $31 million for illegal strike activities directed at Pittston.
>
> The judge said he would wait until Dec. 8 to decide whether to impose additional fines.[20]

The *Charleston Gazette* reported that Judge Glen Williams of the federal district court in Lebanon, Virginia on Thursday, November 2, 1989:

> . . . continued the contempt trial of United Mine Workers Vice President Cecil Roberts Thursday in another effort to avoid action that could undermine pending mediations aimed at ending the Pittston coal strike.

19. *Wall Street Journal,* October 26, 1989.
20. *Williamson Daily News,* October 31, 1989.

Roberts, a principal figure in the nearly 7-month-old strike and the union's primary negotiator after President Richard Trumka, is facing a possible jail term for sitting in front of Pittston Coal Group trucks in defiance of federal and state court orders.

Roberts had acknowledged sitting in front of the trucks in July to protest the lack of movement in earlier federal mediations, but he said he did not feel he did anything wrong.

Defense attorney Scott Mullins requested the continuance of the trial, which had been scheduled for Monday. Senior U. S. Judge Glen Williams granted the continuance.

Former Labor Secretary W. J. Usery has been appointed as a supermediator to try to end the bitter strike. Mullins said a feeling exists that while Usery is working on the negotiations, the courts should try to avoid inflaming the situation.

Usery is preparing for mediations by talking to both sides separately.

Williams' action Thursday was the second time this week a judge has delayed court action in deference to the pending mediations.

While the judges' actions may help keep the union cool, Pittston still is angry about the possibility of a congressional bill that would require them to renew payments to a national industry health benefits fund the company withdrew from in February 1988 when it quit the Bituminous Coal Operators Association.

The bill is aimed at stabilizing the fund, which is more than $50 million in debt.

Pittston officials have said the bill is unconstitutional, would unfairly force the company to subsidize its competitors, and that it would seriously undermine any efforts to end the strike.[21]

Also, the following quoted article from the *Charleston Gazette* contains more information:

> A Virginia judge fined the United Mine Workers, which is striking Pittston Coal, $33.4 million Friday, more than doubling total state penalties assessed the union during the entire 9-month-old walkout.
>
> Among Friday's penalties for more than 150 contempt of court violations was a $4 million fine for the union's four-day takeover of Pittston's huge Moss No. 3 coal preparation plant near Carbo in September.

21. *Charleston Gazette,* November 3, 1989.

Russell County Circuit Court Judge Donald McGlothlin Jr., who already had fined the union nearly $31 million, had delayed further penalties in early November for fear of ruining face-to-face negotiations that were about to start with mediator William Usery.

Those negotiations still are going on, but McGlothlin said "that fear of the court has been allayed" because Usery has been able to keep the parties at the table through 13 sessions.

"Justice delayed is justice denied, and I can't delay too long," McGlothlin said.

"McGlothlin's action came one day after optimistic predictions by Usery's office about next week's round of talks.

"Mr. Usery is hopeful this round of negotiations might be the ones that do it," said Herbert Fishgold, a spokesman for Usery.

UMW spokesman Gene Carroll reacted harshly to McGlothlin's ruling.

"The real miscarriage of justice here is the obscene and what should be unconstitutional level of fines, . . ." Carroll said. "Judge McGlothlin has continued by these excessive fines to do nothing to help settle the dispute with Pittston. Judge McGlothlin's fines allow Pittston to sit back and hope that the fines will destroy the union and bring us to our knees at the bargaining table.

"Companies that have killed workers through occupational hazards have never approached fines like this."

The fines penalize the union for violations of McGlothlin's order banning acts of violence or civil disobedience aimed at slowing Pittston coal traffic or intimidating Pittson employees.

For instance, McGlothlin said, one of the heavier new fines was in reaction to court testimony indicating strike coordinator Marty Hudson led a group of miners wielding baseball bats and clubs in an attack on a coal truck driver. The driver was not hurt.

McGlothlin said he also doubled the maximum daily fine against the union for acts such as blocking coal trucks or the coal plant takover from $500,000 to $1 million.

Usery's negotiations to bring the 1,700 striking miners in Virginia, West Virginia and Kentucky back to the job are scheduled to resume Monday.

The strike started April 5, primarily because the company wanted to scale back medical coverage and had decided to stop paying into a medical benefits fund for retirees managed by the UMW.

The company said it needs the concessions to compete in the international market it serves, while the union said agreeing to the terms of the company's proposed contract would lead to similar demands by the other coal firms.[22]

22. *Charleston Gazette,* December 9, 1989.

The following quoted article from the *Herald-Dispatch* details the status of the $64 million in fines imposed against the union by Judge McGlothlin of Russell County, Virginia:

> A Virginia judge refused Monday to drop $64 million in fines against the United Mine Workers, casting doubt on the future of a tentative settlement reached Jan. 1 between the striking union and Pittston Coal Group Inc.
>
> But Russell County Circuit Judge Donald McGlothlin gave the union an additional 10 days to come up with new arguments in favor of dropping the fines for illegal strike activities.
>
> UMW General Counsel Robert Stropp said the union leadership would reassess its decision to withhold a vote on the tentative contract until all litigation related to the 10-month strike is dissolved.
>
> The union offered to have its 1,700 striking miners in Virginia, West Virginia and Kentucky perform 10,000 hours of community service over 30 days if the fines are waived. McGlothlin dismissed the offer Monday, calling it "an affront to this court."
>
> The judge said dropping the fines would send the wrong message to the community and to future participants in labor disputes.
>
> McGlothlin said it would "convince people . . . that mass violations of the law are a valid means of conducting a strike."
>
> In Charleston, a federal judge Monday dismissed unfair labor practice charges against the UMW on the condition that the strike is settled.
>
> The dismissal order covers only charges against UMW members and locals in West Virginia. The National Labor Relations Board had sought sanctions against the union for alleged unfair labor practices by Locals 1971, 5815 and 8454 against Pittston's Buffalo Mining and Elkay Mining subsidiaries.
>
> The NLRB requested dismissal of its charges and U. S. District Judge Dennis Knapp on Monday said their dismissal "would serve the public good." Charges will be dropped if the union membership ratifies the labor agreement, and picketing also must stop, Knapp said.
>
> McGlothlin, the Virginia judge, said he knew dropping the fines he imposed would make all parties in the dispute happy. Miners would be able to get back to work, state and local governments would be free of the law enforcement burden associated with the often violent strike, and Pittston would be able to return to full production.
>
> Pittston attorney Carl Kindig told the judge Monday a decision to drop the fines would "help bring an end to this dispute in a fashion that will lead to labor peace in this area and ultimately be in the public interest."

But McGlothlin criticized Pittston for asking him to drop the fines after coming to him repeatedly during the strike asking for harsh penalties against the union.

"Now the same plaintiff says everything is fine and we want you to drop everything and wipe the slate clean," he said. "It's an incredible irony in this court's opinion."

"The courts are not pieces of a gameboard to be manipulated," McGlothlin said later.

The fines are the largest obstacle to a final settlement in the strike, union officials said.

Friday, the miners who have been on strike since April, moved a step closer to a vote on returning to work as the NLRB settled lawsuits filed by both the company and the union.

McGlothlin said the NLRB decision to drop its litigation in return for the union's agreement to restrictions on future strike activities had no influence on him. He said his court order restricting strike activities was "found to be totally ineffective."

McGlothlin said there were 263 violent violations of his court order restricting strike activities and 137 non-violent violations.

Meanwhile, U. S. District Judge Glen Williams will decide Tuesday or Wednesday whether to waive nearly $1 million in strike-related fines he assessed against the union, deputy clerk Libby Brown said Monday.

Both judges issued the fines for violations of court orders limiting picket line activity and prohibiting the disruption of Pittston operations.

McGlothlin's decision was seen on the most critical because of the magnitude of the fine total.[23]

The Pittston Settlement

On February 14, 1990, Judge Glen Williams agreed to reduce the amount of fines against the UMW for illegal strike activities from $960,000 to $280,000 and the UMW/Pittston agreement was ratified by the striking Pittston miners on Monday, February 19, 1990. The following article quoted from the *Williamson Daily News* contains more information on the concessions made by each side in the agreement.

> The United Mine Workers union won concessions in health care and pension benefits and Pittston Coal Group gained the

23. *Herald-Dispatch,* February 13, 1990.

ability to run its mines continuously in a proposed strike settlement distributed today to almost 1,700 union members in Virginia, West Virginia and Kentucky.

The proposed four-year pact is to resolve a 10½-month strike between the union and Pittston. The bitter and sporadically violent strike, which gained international attention and national support from union members, began April 5 and could end next week if the UMW's rank-and-file approves the agreement.

The Associated Press obtained a copy of the contract from a source who insisted on anonymity.

UMW President Richard Trumka announced Thursday that union members would vote on the pact Monday, despite $64 million in fines pending in a Virginia state court for illegal strike activity. The agreement would take immediate effect if it is ratified.

More than 2,000 union members packed into a building at the Russell County Fairgrounds this morning to receive copies of the 49-page contract, listen to pep talks by the union's top officers and receive a briefing on the terms.

Trumka choked back tears and he congratulated the miners on what he said was a successful strike.

"I'm proud. I'm proud you did this," Trumka said as he raised the contract in the air to a round of applause.

Pittston and the union reached a tentative agreement on New Year's Eve and have kept the contract a closely held secret while both moved to dismiss lawsuits and fines stemming from the strike.

The proposal in many respects mirrors health and pension benefits contained in the 1988 Bituminous Coal Operators Association contract, the industry's national labor contract.

But it differs from the multi-employer contract in two significant respects: Pittston won the ability to hire contract truck drivers and maintenance workers, so long as union miners are not laid off in the process. And the company would be able to run its mines around-the-clock and seven-days-a-week, except for the Sunday day shift and on Christmas.

Wages were not a major issue in the strike. But miners would receive a cumulative $1.20-an-hour raise at the end of three years. Top-scale pay in selected positions will be $17.52-a-hour.

Health care benefits emerged as a key issue in the contract talks. Pittston dropped out of the multi-employer plan negotiated between the union and the BCOA and proposed payment of 80 percent of the covered medical expenses rather than 100 percent.

The UMW won its demand to maintain the 100 percent medical coverage for employees, pensioners and their families, the same coverage provided by the 1988 national agreement.

The contract has a new health-cost containment provision in which Pittston issues $500 twice-a-year to each miner and pensioner under the age of 65 to cover medical expenses.

After it dropped out the BCOA, Pittston stopped contributing to the multi-employer retirement fund and promised to take care of its own retirees.

But Pittston has agreed to continue participation in the 1974 pension plan, and contribute $10 million to the 1950 benefit fund, the contract said.

"Pension benefits in the new contract are identical to those contained in the 1988 national agreement," according to a contract summary.

The company wanted to be able to contract out work in five job categories but won the limited right in two areas: transportation of coal and repair and maintenance work.

A union source who insisted on anonymity said the company had that right to contract out in those two areas, but the new agreement provides stronger protection for union workers.

The contracting could occur so long as workers employed between Feb. 1, 1988 and April 5, 1989 are not laid off or reassigned, the contract said.

Pittston offered to maintain the pre-strike employment level of 1,695 union employees, but the specific employment level was not addressed in the contract.

"Our understanding is that all operations will go back to work upon ratification of the agreement," said a union source who insisted on anonymity.

But it contained a lengthy section requiring Pittston to hire laid-off union miners for the first four out of every five jobs at Pittston's non-union mines.

Contract miners for Pittston must agree in writing to offer the first 19 of every 20 jobs to laid-off union miners.

And the company would be required, if legally possible, to amend agreements within 90 days with existing contract miners to make similar job offers to laid off union miners.

The company wanted to eliminate a provision that union contracts are not abrogated when a union mine changes hands or is conveyed from one subsidiary to another.

But the settlement retains the successorship provision in the 1984 agreement and contains "strengthened protection for UMWA members against internal company reorganization and shifting of coal lands," the contract said.

Pittston's significant victory was its request to establish round-the-clock operations.[24]

On August 22, 1990, Circuit Judge Donald McGlothlin of Russell County, Virginia rendered his final decision against

24. *Williamson Daily News*, February 16, 1990.

the International Union, UMWA and Clinchfield Coal Company where he originally levied fines against the International Union of $64 million for violations of the laws of the state of Virginia in the union's selective strike against the Pittston Company.

Both the International Union, UMWA, and the Pittston Company requested from Judge McGlothlin that he dismiss these fines in Judge McGlothlin's court on February 12, 1990.

At that time, Judge McGlothlin refused both the union's and the company's requests. However, in his final decision, he dropped $12 million from the $64 million originally assessed against the union, because during the course of the violations of the laws of the state of Virginia, he directed that $12 million of the $64 million be paid to the Pittston Company to cover the damage that was inflicted on the company as a result of the selective strike.

Due to this decision of August 22, 1990, the union must pay approximately $52 million in fines to Dickenson and Russell counties, Virginia, and also to the state of Virginia to cover the reimbursement to the state for the use of the state police in protecting public properties during this dispute.

This decision of Judge McGlothlin will have an adverse effect on the finances of the International Union and also, the union's selective strike fund.

As of December 21, 1989, the union had a cash balance of $100,931,281. Judge McGlothlin's decision of August 22, 1990 will leave the UMW with an approximate balance in its present funds of $48,931,281.

Judge McGlothlin's decision in this case is the largest penalty ever placed upon a labor union.

There are many lawsuits pending against the union by various coal companies throughout the industry as a result of the wildcat strikes that occurred during the selective strike against Pittston. These lawsuits amount to millions of dollars. If the courts render judgment against the UMW, the union will suffer a financial catastrophe.

CHAPTER XXVII

Concluding Observations

The UMWA represented approximately seventy percent of the nation's coal production at the time when Arnold Miller defeated W. A. "Tony" Boyle for the presidency of the union in December of 1972. The nation's coal production represented by the UMW under the presidency of Arnold Miller began a downward trend that continued during the presidency of Sam Church and has also continued during the presidency of Richard Trumka. It reached a new low in 1989 when the UMW represented less than thirty-four percent of the total production of the coal industry. All three of these presidents at the times they were elected to lead the UMW were immature, inexperienced and possessed no qualities of leadership. President Arnold Miller dismissed all of the employees of the union who had been employed during the Boyle administration, and he replaced them with new, young, inexperienced people who had no prior knowledge of the workings of the Mine Workers Union; and, as a result, during Arnold Miller's negotiations of the 1974 contract with the coal industry, he agreed with the operators to take the administration out of the health benefits program of the UMW Welfare and Retirement Fund and turn it over to the signatory companies of the National Bituminous Coal Wage Agreement of 1974. As a result of this concession made by Arnold Miller, the coal companies who were signatory to the agreement had insurance companies underwrite this program and provide benefits at a great savings to the industry, but with severe consequences to the Welfare Fund. This was Arnold Miller's greatest blunder.

Sam Church, who succeeded Miller as a result of Miller's illness, was president for a short time. He negotiated the 1981 National Agreement which provided for daily wage increases of:

$1.20 —	June 1981;
.95 —	June 1982;
1.00 —	1983, and
.45 —	1984

Also, for the first time in the history of collective bargaining, he was able to have written into this national agreement that the miners widows would receive a pension of $100 per month. This was quite an accomplishment and stands out as the most outstanding contract negotiated, as between Arnold Miller, Sam Church and Richard Trumka.

In 1982, Sam Church announced his candidacy for president of the Mine Workers Union for a five-year term, and Richard Trumka challenged Sam Church. In December of 1982, Richard Trumka was elected president of the UMW Union by an overwhelming majority of the membership of the union. Trumka negotiated the 1984 agreement that expired on January 31, 1988, and he negotiated the 1988 agreement, a five-year contract, that became effective February 1, 1988 running through February 1, 1993. During the negotiations for the 1988 agreement, Trumka stated publicly to the media that the union would not take any backward steps and would not make any concessions during the negotiations. However, after the contract was signed and put into effect February 1, 1988, Trumka made a major blunder by concurring with the new agreement which called for the elimination of the tonnage rate paid into the Health and Benefits Fund and substituting in its place an increase from $1.11 to $1.83 *per hour worked* by each individual miner. This amounts to only $14.64 per day for eight hours for each working miner as a contribution to the Welfare and Retirement Fund, at a time when coal tonnages have increased and the number of employees has decreased. As a result of this change, the Health and Retirement Fund has not been able to meet its obligations because of the reduced contributions into the Welfare and Retirement Fund.

As of October 31, 1989, the 1950 Health and Benefits Fund reached a new low by being $54 million in the red in its accounts.

The UMW Welfare and Retirement Fund audit by Peat Marwick Main & Co. of Washington, D.C. covering the period from July 1, 1987 through June 30, 1988 shows the balance of the 1950 Benefit Trust Fund was in the red $18,086,000. In a news item of August 3, 1989, carried in the local paper, it was re-

reported that the Health and Benefit Fund had a deficit of $37 million since June 30, 1988. In the past year, the fund continued to show a deficit of approximately $3 million per month. This Health and Benefit Fund cannot continue on its present course.

Trumka requested that the signatory companies make up this deficit, and they all rejected his request. Also, the trustees of the Welfare and Retirement Fund brought suit in the Federal District Appeals Court of Washington, D.C. to have the signatory companies increase the hourly rate so that this indebtedness could be cleared up. On April 5, 1989, the three-man District Appeals Court panel ruled unanimously that the National Agreement did not permit the trustees to increase the hourly rate to make up this deficit as had been the case under earlier agreements when John L. Lewis was a trustee, and Trumka found himself in a dilemma. Letters have been forwarded to some of the miners who were recipients of medical attention from doctors and hospitals stating that they have billed the Health and Benefits Fund several times in the last six months and were unable to receive any payment for the services the individual miner received. The letters asked the miners to make some arrangement to make payment to help liquidate their obligation with the hospital or doctor, as the case may be. As a result of this activity, Mr. Trumka is in serious trouble.

A temporary expedient has just been announced: "By an amendment to the BCOA-UMWA current labor agreement, Joseph Brennan, on behalf of the BCOA, and Richard Trumka on behalf of the UMWA have agreed to a transfer of funds from the 1974 pension trust to the 1950 Health and Benefits trust." This raises the question as to what effect this will have in the longer term on the pension plan, since operator contributions are now based on man-hours rather than tonnage.

Serious problems now exist within the once great UMW Union because of departures from five principles held to be inviolate by John L. Lewis during his tenure:

1. No strike fund—miners should be well paid for their work and they each should personally set something aside to sustain them and their families during strike emergencies.
2. Sound fiscal policies—build up the union's treasury and assets.
3. Organize the unorganized—above all, the miners so that once a contract has expired, all mining ceases in the industry—no general policy on selective strikes.
4. Loyalty of the organization to its members who carry on its activities and in so doing suffer injury to themselves and their families.
5. Independence and integrity of the union.

As to the *strike fund*, in the 1983 convention, just after he was elected for his first term, Trumka arranged for a change in the union constitution which would grant him full authority to call selective strikes and to assess working miners up to two and one-half percent of their wages for a selective strike fund to be capped at $70,000,000.00. In a special convention in Atlanta in 1986, the cap was removed. Trumka is now reportedly urging an increase in the assessment of five percent. It is also alleged that in the forthcoming UMW Constitutional Convention to be held at the Fountainebleau Hotel in Miami, Florida in September 1990, Trumka will seek to increase the dues of the miner from $43.50 per month to $46.81 (dues were $12.00 under Miller). It will be interesting to note whether or not the delegates to this convention will continue this increase into the Selective Strike Fund. There is no record of any labor organization in this country that has ever extracted such large sums of money from its membership as the UMWA.

For the past six years, Trumka has extracted from the pay envelopes of the working miner approximately $300 per month for a strike fund and according to his secretary-treasurer's testimony before Judge McGlothlin of Russell County, Virginia, the average monthly collection from this special assessment is $500,000 per month. Since the strike against Pittston which began April 5, 1989, the secretary-treasurer stated it was costing the union approximately $4,000,000 a month in strike benefits to those who are on selective strike. The average mine workers pay is approximately $600 per week, but when you have him on selective strike, his income drops to $225 per week which is paid by the union. This situation has existed since the A. T. Massey strike four years ago and when you talk to the average miner, he has already started to grumble that he can't make it on the $225 per week paid to him by the union and take care of his other obligations. This unrest brewing within the union by the active members of the union raises the question of whether the selective strike weapon will be effective in the future.

As to *sound fiscal policies*, during the six-year period from 1983 through 1989, the union had total receipts of $675,664,236 but expenditures of $575,112,324. In addition, however, assets accumulated largely through John L. Lewis's administration have been disposed of, including considerable real estate and a seventy-six percent ownership of the National Bank of Washington.

In accordance with the Labor Organization Annual Report,

Form LM-2, filed by the UMWA with the U.S. Department of Labor, the following is a listing of receipts and expenditures for the International Union from 1983 through 1989:

Receipts		Expenditures	
1983:	$ 18,968,258	1983:	$ 19,312,250
1984:	$114,963,508	1984:	$ 41,003,829
1985:	$125,577,218	1985:	$ 66,406,846
1986:	$ 39,337,761	1986:	$101,724,236
1987:	$160,641,202	1987:	$129,835,823
1988:	$ 93,568,618	1988:	$ 91,805,971
1989:	$122,607,671	1989:	$125,023,369
Grand Total	$675,664,236		$575,112,324

On *organizational* efforts, no longer is the UMWA the pillar of American unionization—let alone the pillar of labor representation in the coal industry.

Since the advent of the Miners for Democracy regime, under the leadership of Arnold Miller beginning in 1973, and continuing through November 1979, the decline in UMW-produced coal has been as follows:

Coal Production Trends, NBCWA and Overall U.S.
(Million Short Tons)

Year	Total Production	NBCWA*	% of Total
1973	591.7	392.3	66.3
1974**	603.4	367.4	60.9
1975	648.4	381.4	58.8
1976	678.7	369.5	54.4
1977	691.3	330.8	47.8
1978**	665.1	273.1	41.1
1979	776.3	356.9	46.0

*National Bituminous Coal Wage Agreement Mines (UMWA)
**Coal strike years
(Figures are rounded)
SOURCE: Bituminous Coal Operators Association[1]

This is an approximate twenty percent decline in union-produced coal during this period.

Vice-President Sam Church succeeded Arnold Miller on November 16, 1979 as president of the union and unionized produced coal in the industry continued its downward scale.

1. National Coal Association, *1983 Facts About Coal*, p. 18.

Year	Total Production	NBCWA*	% of Total
1980	823.6	364.9	44.3
1981**	818.4	316.7	38.7
1982	829.2	351.5	42.4

*National Bituminous Coal Wage Agreement Mines (UMWA)
**Coal strike year
(Figures are rounded)
SOURCE: Bituminous Coal Operators Association[2]

Also, unionized coal production has continued its downward scale under the leadership of Richard Trumka, who became president of the UMWA in December of 1982.

Bituminous
Coal Production Trends, NBCWA and Overall U.S.
(Million Short Tons)

Year	Total Production	NBCWA*	% of Total
1983	778.0	312.0	40.1
1984	891.7	344.2	38.6
1985	878.9	320.8	36.5
1986	886.0	314.5	35.5

*National Bituminous Coal Wage Agreement Mines (UMWA)
(Figures are rounded)
SOURCE: Bituminous Coal Operators Association[3]

According to the latest information available from the National Coal Association, the following is the coal production figures for 1987, 1988 and 1989:

1987 Total Production:	901,626,000
UMWA-Produced Tonnage:	320,000,000 (estimated)
Percentage of Total:	35.5%

1988 Total Production:	958,912,000
UMWA-Produced Tonnage:	326,000,000 (estimated)
Percentage of Total:	34.0%

1989 Total Production:	979,578,000
UMWA-Produced Tonnage:	330,000,000 (estimated)
Percentage of Total:	33.7%

2. National Coal Association, *1983 Facts About Coal*, p. 18.
3. National Coal Association, *Facts About Coal 1988*, p. 30.

As for the union standing by its loyal members who get in trouble doing the union's work, recent examples belie the old UMWA tradition.

The union had been extracting from the working miners pay envelopes for the past three years one and one-half percent of their monthly wages averaging $300 or more per month or approximately $3,600 per year per miner into the union's strike fund. This was beginning to place the miner in a hardship position because these assessments have been in effect since the selective strike policy of Mr. Trumka was instituted. Also, they had miners running all over several states soliciting financial aid from members of other unions and from their national unions. They were taking in millions of dollars and not giving an accounting of these funds. They were paying the pickets where the selective strike was being instituted approximately $225 per week and were using all kinds of methods to pressure the retirees and those miners who were receiving health benefits to man the picket lines or have their benefits placed in jeopardy. The miners who are manning the picket lines have no protection from the union in the event they are shot or killed while serving on the picket line. Their families would have to appeal for assistance for their livelihood from the local welfare agencies. A good example of this is in the selective strike against A. T. Massey Coal Company where one of the truck drivers who was hauling coal to the Massey cleaning plant in Pike County, Kentucky was killed. Initially, the four individual mine workers who were indicted and prosecuted were receiving $225 per week for being on strike. After their convictions and sentencing, their families checks from the union were discontinued. The following article quoted from the *Williamson Daily News* contains more information:

> The wife of a union miner convicted in the sniper death of a non-union coal truck driver said she believes the union owes her family and the families of others convicted in the incident.
>
> "Our husbands gave up their lives for the union, and the families gave them up to the union. I think they owe us something," Barbara Thornsbury said. "We are on our own. We were forced to do something we were not prepared to do."
>
> Thornsbury's husband, David, and four others were convicted in the sniper shooting death of Hayes West, a non-union truck driver.
>
> West, 35, was shot May 29, 1985 as he hauled coal on U. S. 119 on Colburn Mountain in Pike County near the Kentucky-West Virginia border. Another man, Roy Lee Johnson, of Speight, was injured in the incident.

West was the victim of an ambush-style shooting that occurred as he was leading a three-truck convoy across the mountain from a coal preparation plant at Sharondale to a mine site at Canada.

According to Kentucky State Police reports, West's truck was struck by 21 rounds. Two of those rounds struck the driver's window and door.

The attack came during the bitter United Mine Workers/A. T. Massey Coal Co. strike. West was hauling coal from Samoyed Coal Corp., which was also picketed by union members.

David Thornsbury, along with the others convicted in the incident, was being paid benefits through the union's Strike Fund, which supplied money, prescription drugs and hospitalization insurance to striking union families.

Thornsbury's first cousin, Donnie, also was convicted of the shooting. He was president of UMW Local 2496 at Canada.

Barbara Thornsbury's Strike Fund benefits were eliminated in April of this year, after her husband's last-ditch appeal was denied.

Federal prosecutors asked a U. S. appeals court in Cincinnati in late September 1988 to uphold the conviction of the four union members who were charged in the ambush.

The federal lawyers defended the prosecutions' interpretation of a federal statute used to convict the men, which held that a truck driven by the slain driver was part of interstate commerce even though it remained in Kentucky.

A defense lawyer argued, however, (that it) is incorrect and the conviction of the four defendents should be thrown out.

A three-judge panel of the 6th U. S. Circuit Court of Appeals took the case under review. It later upheld the convictions. She said she received a phone call by a local union member, telling her that her husband's benefits were being cut off, shortly after his appeal was denied.

"The guys from the international (union) were to [*sic*] chicken to call me, so they had one of the local boys do it," she said.

She said she was told her benefits were being denied because her husband's last appeal had been denied. She said she never got any written information from the union.

"I have been informed of very little," she said.

She said Roger Hammack, a UMW official in Washington, D. C. confirmed that she had been cut off from the Strike Fund. She said that is all she has heard from the union.

Hammack was contacted by the *Daily News,* but said "No comment, period" to all questions asked.

Thornsbury added that all other families involved in the incident have also been removed from the Strike Fund, but before she was. "Why that happened, I just don't know," she said.

The fund provided $200 per week, plus prescription drug and hospital coverage. She said she applied for Social Security benefits, but was turned down. She said the medical coverage was very important to her, because she has severe allergies and other illnesses that required medication.

She said that the incident has totally reversed her life, which was once "very secure."

"We had the entire family unit, then all of a sudden, it was torn apart. It was just incredible."

She said she has written two letters to Richard Trumka, UMW President, but has received no reply from him.[4]

On August 19, 1989 Roy Blankenship, an employee of Elkay Mining Company, a subsidiary of the Pittston Company, was shot while he was picketing Slab Fork Hollow Coal Company which is a subsidiary of A. T. Massey Coal Company. At the time Blankenship was wounded, the UMW commented that Blankenship was a hero due to the fact that he was wounded while supporting the cause of the UMW. In a newspaper interview on May 30, 1990 Blankenship was very disenchanted and critical of the manner in which the District 17 officers, particularly Richard Trumka, president of the International Union; Cecil Roberts, vice-president; and John Banovic, secretary-treasurer, handled his particular case when he applied for assistance for his recovery and disability as a result of his being wounded. The following news item appeared in the *Charleston Daily Mail* covering his complete interview with the paper regarding his disappointment with the union:

Roy Blankenship was just another humble coal miner when he joined the United Mine Workers union 25 years ago.

This past year changed all that. Fighting for the UMW, Blankenship has risked his job, risked being arrested and risked his life.

He took a bullet in his right arm. The bullet, allegedly fired by A. T. Massey Coal Co. contractors, hit him during a week-long spate of violence between the union and the company last August in Slab Fork Hollow.

Quickly heralded as a hero by the union, Blankenship stood as a rallying symbol of the union's cause against corporate greed and government inaction.

But Blankenship, 46, speaks bitterly of the way the UMW has treated him in recent months. He is especially critical of the UMW leadership under President Richard Trumka and Vice President Cecil Roberts.

4. *Williamson Daily News*, June 7, 1989.

"The international was supposed to help me, but it hasn't done anything," Blankenship, vice president of UMW Local 1971, said late Friday from his Amherstdale home. "I believe this is the worst administration we've had in my 25 years in the union."

Blankenship started receiving UMW strike fund benefits of $225 a week and health insurance when the union struck The Pittston Co. in April 1989. After being shot in August, it was doubtful if he would ever be able to work again. His arm is paralyzed, but he is recovering and may be able to work again. He said the union promised him full benefits when the strike ended.

UMW officials in Washington, D.C., said Tuesday they are doing everything they can to help Blankenship. Before considering alternatives, they are attempting to get him sickness and accident benefits from Pittston, but the case has been mired in bureaucracy.

Blankenship said he has heard that before—the company already rejected his application for benefits. But Marty Hudson, a UMW organizer, said Blankenship should still be able to collect the benefits once negotiations with Pittston conclude.

"Roy is a good guy. It's a terrible situation what happened to him," Hudson said. "We are going to help Roy."

Privately, some other union members think Blankenship is turning his benefits issue into a political issue. He says that's not true.

The short, stocky miner is running for a field representative position in the UMW's troubled Logan subdistrict in a few weeks. UMW international and district officials have all but taken over the subdistrict in the wake of numerous complaints of vote fraud last year.

A field representative must do a considerable amount of paperwork and arbitration of grievances. The job pays more than $40,000 a year. Blankenship says despite his injury, he can do the work and can write again.

He claims the international leadership is not supporting his candidacy, but that is not why he is upset.

A number of union stalwarts in Logan County, including ousted district board member Danny Wells, also have been critical of the union leadership. UMW leaders admit the criticism has almost become fashionable in Logan, but vow to continue the cleanup of alleged fraud in the subdistrict.

When Local 1971 was voting to ratify the UMW's new contract with Pittston last February, Blankenship said Vice President Roberts approached him and said the international would take care of him.

At the time, the union leadership lobbied heavily for ratification and, in the process, promised to put 13 suspended Pittston employees on the international payroll—at working wage scales—until they were called back.

Blankenship said union leaders indicated he would receive a package similar to the one being offered to the suspended miners. Instead, he only collected the $225 a week from the strike fund. If he had received a similar package, he would have been paid at least an additional $500 more a week.

"It makes me mad," he said. "It makes a lot of our people mad. Some of the guys in (Local) 1971 are writing a letter to the president, vice president and secretary-treasurer and taking donations to give me something to live on.

"They know how hard it is. They know you can't raise a family on $450 every two weeks."

A husband with a daughter in college, Blankenship said he "raised hell" with the union at both the international and district levels, but has not received answers. So he decided to make his case public—a move not usually received well by union leaders.

"I don't care what they do," he said of the potential reprisals. "They can't do anything more than they've already done."

Article 10 of the UMW constitution allows members to freely express views about the policies and business of the union, but it also mandates loyalty to the union.

Hudson, a longtime union official who hails from Boone County, said the benefits issue should be resolved in a matter of weeks.

Bruce Bratten, a top aide to Trumka, is working on the case personally, Hudson said. Blankenship may also be entitled to back benefits, he added.

Although he said he likes Hudson personally, Blankenship contends that the union leadership doesn't care about him anymore. He claims the reason is because he is not a part of "the clique" that follows the Trumka administration.

"I don't kiss up after the international or district," he said. "If you're not in the clique, they don't have much use for you."

Although he continues to campaign for the field representative spot, Blankenship said it won't be the end of the world if he loses. He was actually called back to work by Pittston's Elkay Mining Co. subsidiary in March, but couldn't respond because of the gunshot wound complications.

His doctors recently told him there is a possibility he could work again.

In the meantime, six employees of Mate Creek Trucking Inc. are facing a possible trial in Logan Circuit Court on attempted murder charges for allegedly shooting Blankenship on August 19.

But the trial has not been scheduled. The state Supreme Court recently agreed to hear arguments from the Mate Creek employees, who are seeking to have their indictments dismissed. They claim there were technical errors made by the prosecution and the grand jury and that there is insufficient evidence to support the crimes theyare charged with.

Mate Creek and Con-Serv Inc. are Massey's contractors at Slab Fork.[5]

This is another factual complaint which shows the disregard by the district officers and international officers in not providing Blankenship with any assistance for the injury he received on the picket line at this coal company. In the forepart of this chapter, we have noted that the four women whose husbands were tried and convicted under the Hobbs Act in the death of Hayes West, who was killed by ambush, had a statement published in the *Williamson Daily News* that their weekly benefits paid out of the union's strike fund had been terminated. They were notified of this action by the subdistrict officers of the union and the reason they were given for the termination of their benefits was due to the union losing the appeal of these four members of the union. This action by the union became effective in April of 1989.

Both of these incidents have been well publicized, and the members of the union who are called upon to picket mines, as they did in both of these cases, should think long and hard before they subject themselves to the exposure of picketing in support of the union's cause and make a sacrifice in behalf of the union. When the union is asked to support them, as in the case of Blankenship and the wives of the four men who were tried and convicted under the Hobbs Act, they receive no protection from the union in the event they get into trouble on the picket line.

In 1921 during the so-called Blair Mountain War, there were three individuals who were ambushed and killed on Blair Mountain by someone in the miners ranks. Since these three individuals were there supporting the cause of the coal operators, the operators in the Logan area paid for the burial of these three individuals. Also, the operators paid $30 per month to these individuals' widows from 1921 until 1964 when the last surviving widow died. These widows received this money up until the time of their deaths covering a period of forty-three years of $30 per month for each of the widows. It is to be noted that the operators are more humanitarian in the cause in which their employees are more dedicated to serve the coal operators and have far greater protection as shown in this case.

There has been a lot of stories written about the war on Blair Mountain where hundreds of people were killed. This is all un-

5. *Charleston Daily Mail*, May 30, 1990.

true. There were only three individuals that were supporting the cause of the operators who lost their lives.

As for the independence and integrity of the UMWA as a union:

According to a press release on Thursday, October 5, 1989, "The on again/off again relationship between the United Mine Workers of America and the AFL-CIO is about to change once again.

"The International Executive Board of the UMW voted Wednesday to try to rejoin the AFL-CIO, the labor federation that the miners' union left in 1947 in a disagreement about the Taft-Hartley Act."[6]

Mr. Lewis's position in the disagreement between the UMW and the AFL over the Taft-Hartley law was that the AFL and its affiliates should ignore the Taft-Hartley law and should not avail themselves of any provision of this anti-labor law. He asked the executive council of the AFL and other labor unions to ignore the law because the law required that the officers of labor unions were required to sign non-Communist affidavits. Mr. Lewis took the position that it would automatically classify labor leaders as second-class citizens. The AFL refused to go along with the request of Mr. Lewis, and he withdrew from the AFL and the UMW union became an independent union for the past forty-two years.

Mr. Trumka taking the UMW back into the AFL-CIO makes the UMW one of the smallest major unions within the AFL-CIO, with approximately 100,000 active members. We cannot see how this affiliation will help the membership of the UMW other than to say that they are affiliated with the AFL-CIO. This is just another gimmick that Mr. Trumka has come up with in order to try and salvage his presidency.

The miners union was founded on January 25, 1890 in Columbus, Ohio. On January 25, 1940, the UMWA held its golden anniversary convention in the same building that the union was founded in 1890. President John L. Lewis, Vice-President Phillip Murray and Secretary-Treasurer Tom Kennedy extolled the pride and successes the miners union had accomplished during its first fifty years of existence. At that convention in 1940, the membership of the UMWA had risen to the highest level in its history. The international officers, over a national radio hook-up on NBC, reported that the union had reached a goal of approximately 600,000 members. Its finances

6. *Charleston Gazette*, October 4, 1989.

and prestige also reached the highest of any labor union on the North American continent. There were more than 3,000 delegates attending that convention, representing ninety-eight percent of the coal industry, and the delegates were overwhelmed and jubilant at the successes and accomplishments of the first fifty years of the UMW's existence.

The UMW Union has now celebrated its One Hundredth anniversary. What a contrast from 1940 when it celebrated its first fifty years of existence. The membership of the UMWA from 1940 to January 25, 1990 has declined to the point that the union only represents approximately thirty-four percent of the coal produced in the coal industry or in other words, sixty-six percent of the coal produced on the North American continent is produced by nonunion labor. This is a decided change from 1940 when the union had mine workers contracts with ninety-eight percent of the operating coal companies in this country. The membership decline began in 1973 under the leadership of Arnold Miller and continued its downward trend until 1983 when Mr. Trumka assumed the presidency of the UMWA. Since then, under the leadership of Mr. Trumka, this decline has escalated even further. Without the income from the assessment received since 1983 into the strike fund, the mine workers union finances would be in a shambles, because during Mr. Trumka's tenure of office, he has sold off all of the fixed assets and property that the union had accumulated.

Under normal circumstances any organization such as the UMW with the history of success in its first fifty years would continue to improve its standing. This is not true of the UMWA today. In fact during the past sixteen years, there is no major labor organization in America that has declined to such a low ebb, both in its membership and prestige, as did the UMWA.

In sizing up the leadership of the mine workers union since W. A. Boyle's presidency in 1972, in my view, Sam Church would be number one, even though his tenure of office was short. Number two would be Arnold Miller and number three would be Richard Trumka. It will be interesting to see in the next eighteen months how Mr. Trumka is going to handle the many problems he is confronted with today. Time will tell the future of the UMWA.

Following this chapter, we have inserted the accomplishments of John L. Lewis during his forty-year reign as president in addition to some of his outstanding speeches during this period, and a comparison of the three major agreements that have been signed by the UMW under the selective strike policy instituted by President Richard Trumka.

Also, we have included a report on the Selective Strike Fund by the International Union, UMWA and the decision of Judge McGlothlin rendered on August 22, 1990 in the Russell County, Virginia Circuit Court.

APPENDIX I

Accomplishments of John L. Lewis 1920-1960

1920 Elected to five-year term as president of UMWA

1924 Negotiated Jacksonville Agreement—February 18, 1924 to April 1, 1927

*Contained highest wage scale in the Central Competitive Field

1928 Policy Committee of International Union negotiated $6.10 basic daily wage in Illinois and Indiana; Southwestern District $5.00 day rate and Northwestern District $6.72 basic daily rate

1933 (First Appalachian Wage Agreement) Covered seventy percent of National Tonnage—October 2, 1933 to April 1, 1934

*$5.00 daily wage for mines north of the Ohio River

*$4.60 daily wage for mines south of the Ohio River

*Basic loading rate per ton $.422

*Hours of labor shall not exceed forty hours per week, eight hours per day

Provision for accurate weighing of coal mined on a tonnage basis

*Granted miners the right to select own checkweighman

*Outlawed company scrip or token money, provided that wages be paid semi-monthly in lawful money or check without discount

*Provided that mine workers could not be compelled to live in company houses or trade at company stores as a condition of employment

*Barred sixteen year olds from working in a mine

*Provided divisional labor boards to settle disputes

1934 UMWA moved International Headquarters from Indianapolis, Indiana to Tower Building in Washington, D.C.

1934 Appalachian Wage Agreement—April 1, 1934 to March 31, 1935

*$5.00 daily wage in the north; $4.60 daily wage in the south

*Workday reduced from eight to seven hours for mines north of the Ohio River

*Basic loading rate per ton $.502

*Check off for union dues and assessments

*Complete recognition of the union

1935 Appalachian Wage Agreement—October 1, 1935 to March 31, 1937

*Basic loading rate per ton $.582

1935 November 9, 1935—Formation of the Committee for Industrial Organization

1937 Appalachian Wage Agreement—April 1, 1937 to March 31, 1939

*$6.00 daily wage rate in the north

*$5.60 daily wage rate in the south

*Seven-hour workday (universal)

*Creation of Mechanized Mining Commission

*Basic loading rate per ton $.662

*Overtime at time and one-half

1937 December, 1937 International Union Headquarters moved into University Club on 15th and Eye streets, Washington, D.C.

1939 Appalachian Wage Agreement—May 12, 1939 to March 31, 1941

*No wage increases

*Union shop clause

1941 Appalachian Wage Agreement—April 1, 1941 to March 31, 1943

*Universal basic daily wage of $7.00

*Elimination of $.40 wage differential. Day rate increased to $1.00 in the north, $1.40 in the south

*Annual vacation payment of $20

*Union shop universal in commercial mines

*Seniority based on length of service

*Basic loading rate per ton $.772

1943 February 10, 1943—memorandum executed in reference to six-day work week. Six-day week authorized with time and one-half in excess of thirty-five hours per week

1943 June 18, 1943—War Labor Board Directive eliminating lamp and smithing charges and providing that management furnish tools free of cost

*Increase in vacation bonus to $50.00. Vacation period waived due to national emergency

1943 Ickes-Lewis Agreement—December 17, 1943 to March 31, 1945

*Portal-to-portal pay

*Eight hours work, fifteen minutes travel time and fifteen minutes lunch time

*Wage increase of $1.50 per day

1945 National Bituminous Coal Wage Agreement (first nationwide contract)—April 11, 1945 to April 1, 1946

*New overtime arrangement increased pay of $7-a-day inside men from $57.06 to $63.50 a week

*Equalization of rates in unit mechanized production gave drillers and shooters a larger share of the benefits of mechanization

*Shift differential payments

*Nine-hour day (seven hours at regular pay, one hour at time and one-half, and one hour at $1.50)

*Piece workers receive one-ninth of daily earnings as pay for travel time

*Paid 15-minute lunch period

*Increase in vacation payment from $50.00 to $75.00

1946 Krug-Lewis Agreement—May 29, 1946

*Flat increase per day $1.85

*Vacation payment $100.00

*Pro-rata vacation payments

*Establishment of Federal Mine Safety Code

*Establishment of UMWA Welfare and Retirement Fund. Operators to contribute $.05 per ton on all coal produced for use or sale

*Creation of a Medical and Sanitary Survey of the Bituminous Coal Industry

*Creation of a Medical and Hospital Fund financed from deductions from miners pay

*Holidays, when worked, paid at time and one-half, or rate and one-half

1947 Fairless-Humphries-Lewis Agreement—July 1, 1947 to June 30, 1948

*Mine Safety Code

*Increase in royalty payments to Welfare and Retirement Fund from $.05 to $.10 a ton

*Working day reduced to seven and one-quarter hours for outside men and working day reduced for inside workers from nine to eight hours a day, portal-to-portal

*Basic daily wage increased from $11.85 to $13.05 (increase $3.05-$1.85 plus $1.20 per day on rates effective March 31, 1946)

*Paid lunch period increased from fifteen minutes to thirty minutes

*Insertion of "Willing and Able Clause"

1948 National Bituminous Coal Wage Agreement—July 1, 1948 to June 30, 1949

*Increase in wages for each union employee of a flat $1.00 per day

*Increase in payment to Welfare and Retirement Fund of $.10 per ton, making the total payment $.20 per ton

*Basic wages $14.05 a day (increase $4.05-$3.05 plus $1.00 per day on rates effective March 31, 1946)

1949 Purchase of seventy-six percent of the outstanding common stock of the National Bank of Washington for controlling interest by the union. Price paid: $25,000,000

1950 Love-Lewis Agreement—March 5, 1950 to June 30, 1952

*Welfare Fund payments increased to $.30 per ton

*"Willing and Able Clause" eliminated

*Wage increase of $.70 per day, bringing basic wage to $14.75 a day (increase $4.75—$4.05 plus $.75 per day on rates effective March 31, 1946

*A new Welfare Board of Trustees appointed with Mr. Lewis, chairman, Welfare Fund Director Josephine Roche, Neutral Trustee, and Charles Owens of New York, Operator Trustee

1950 Purchase of lot adjacent to UMW Building which extended from Eye Street to K Street and purchase of five-story building on the corner of 15th and K streets where the offices of the UMW Journal, Contract Enforcement Division and Mine Safety Department were moved to

1950 Mr. Lewis assisted in establishment of the Bituminous Coal Operators Association. Harry Moses, formerly in charge of the coal division of U.S. Steel, named first president of BCOA on July 1, 1950

(NO STRIKES IN INDUSTRY FROM 1951 TO 1956.)

1951 Moses-Lewis Agreement—February 1, 1951 to March 31, 1952 (Amendment to 1950 Agreement)

*Wage increase of $6.35 ($4.75 plus $1.60)

1952 Moses-Lewis Agreement—October 1, 1952 to September 30, 1953 (Amendment to 1950 Agreement)

*Wage increase $8.25 ($6.35 plus $1.90)

*Ten cents per ton increase on Welfare Fund—total $.40 per ton

*Seniority provisions

1955 Moses-Lewis Agreement—September 1, 1955 to August 31, 1956 (Amendment to 1950 Agreement)

*Wage increase $9.45 ($8.25 plus $1.20); April 1, 1956, increase to $10.25 ($9.45 plus $.80)

*Vacation pay increased from $100.00 to $140.00 annually

*Saturday work at time and one-half and Sunday work at double time, as such (pay for the two premium days to be calculated for the period beginning 12:01 a.m. and ending 12:00 p.m. respectively)

1956 National Bituminous Coal Wage Agreement—October 1, 1956 to September 30, 1957 (Amendment to 1950 Agreement)

*Wage increase $11.45, effective October 1, 1956; increase $12.25, effective April 1, 1957

*Annual vacation of fourteen calendar days

*Vacation pay increased to $180

1956 Dedication of ten Appalachian Regional hospitals

1958 National Bituminous Coal Wage Agreement—December 1, 1958 to November 30, 1959 (Amendment to 1950 Agreement. Negotiated between Mr. Lewis and Edward G. Fox who succeeded Harry Moses as president of BCOA)

*Wage increase, effective January 1, 1959, to $13.45; wage increase, effective April 1, 1959, to $14.25

*Provision for Protective Wage Clause

*Vacation pay increased to $200.00

1959 Mr. Lewis instrumental in forming together with George H. Love, chairman of the board, Consolidation Coal Company, the National Coal Policy Conference on February 25, 1959

1960 Mr. Lewis retired as president of the UMWA on January 14, 1960. Thomas Kennedy, vice-president of the UMWA, succeeded Mr. Lewis as president of the UMWA

APPENDIX II

Public Relation Speeches of John L. Lewis 1937 to 1947

Labor and the Nation

"Out of the agony and travail of economic America the Committee for Industrial Organization was born. To millions of Americans, exploited without stint by corporate industry and socially debased beyond the understanding of the fortunate, its coming was as welcome as the dawn to the night-watcher. To a lesser group of Americans, infinitely more fortunately situated, blessed with larger quantities of the world's goods and insolent in their assumption of privilege, its coming was heralded as a harbinger of ill, sinister of purpose, of unclean methods and non-virtuous objectives.

"But the Committee for Industrial Organization is here. It is now and henceforth a definite instrumentality, destined greatly to influence the lives of our people and the internal and external course of the republic.

"This is true only because the purpose and objectives of the Committee for Industrial Organization find economic, social, political and moral justification in the hearts of the millions who are its members and the millions more who support it. The organization and constant onward sweep of this movement exemplifies the resentment of the many toward the selfishness, greed, and the neglect of the few. The workers of the nation were tired of waiting for corporate industry to right their economic wrongs, to alleviate their social agony and to grant them their political rights. Despairing of fair treatment, they resolved to do something for themselves. They, therefore, have organized a new labor movement, conceived within the principles of the national bill of rights and committed to the proposition that the workers are free to assemble in their own forums, voice their own grievances, declare their own hopes and contract on even terms with modern industry for the sale of their only material possession—their labor.

"The Committee for Industrial Organization has a numerical enrollment of 3,718,000 members. It has thirty-two affiliated national and international unions. Of this number eleven unions account for 2,765,000 members. This group is organized in the textile, auto, garment, lumber, rubber, electrical manufacturing, power, steel, coal and transport industries. The remaining membership exists in the maritime, oil production and refining, shipbuilding, leather, chemical, retail, meat packing, vegetable canning, metalliferous mining, miscellaneous manufacturing, agricultural labor, and service and miscellaneous industries. Some 200,000 workers are organized into 507 chartered local units not yet attached to a national industrial union.

"This record bespeaks progress. It is a development without precedent in our own country. Some of this work was accomplished with the enlightened cooperation or the tolerant acquiescence of employers who recognized that a new labor movement was being forged and who were not disposed, in any event, to flout the law of the land. On the other hand, much of this progress was made in the face of violent and deadly opposition which reached its climax in the slaughter of workers paralleling the massacres of Ludlow and Homestead.

"In the steel industry the corporations generally have accepted collective bargaining and negotiated wage agreements with the Committee for Industrial Organization. Eighty-five percent of the industry is thus under contract and a peaceful relationship exists between the management and the workers. Written wage contracts have been negotiated with 399 steel companies covering 510,000 men. One thousand thirty-one local lodges in 700 communities have been organized.

"Five of the corporations in the steel industry elected to resist collective bargaining and undertook to destroy the steel workers' union. These companies filled their plants with industrial spies, assembled depots of guns and gas bombs, established barricades, controlled their communities with armed thugs, leased the police power of cities and mobilized the military power of a state to guard them against the intrusion of collective bargaining within their plants.

"During this strike eighteen steel workers were either shot to death or had their brains clubbed out by police, or armed thugs in the pay of the steel companies. In Chicago, Mayor Kelly's police force was successful in killing ten strikers before they could escape the fury of police, shooting eight of them in the back. One hundred sixty strikers were maimed and injured by police clubs, riot guns and gas bombs and were hospitalized. Hundreds of strikers were arrested, jailed, treated with brutality while incarcerated and harassed by succeeding litigation. None but strikers were murdered, gassed, injured, jailed or maltreated. No one had to die except the workers who were standing for the right guaranteed them by the congress and written in the law.

"Governor Davey, of Ohio, successful in the last election because of his reiterated promises of fair treatment to labor used the military power of the commonwealth on the side of the Republic Steel Company and the Youngstown Sheet and Tube Company. Nearly half of the staggering military expenditure incident to the crushing of this strike in Ohio was borne by the federal government through the allocation of financial aid to the military establishment of the state.

"The steel workers have now buried their dead, while the widows weep and watch their orphaned children become objects of public charity. The murder of these unarmed men has never been publicly rebuked by any authoritative officer of the state or federal government. Some of them in extenuation, plead lack of jurisdiction, but murder as a crime against the moral code can always be rebuked without regard to the niceties of legalistic jurisdiction by those who profess to be the keepers of the public conscience.

"Shortly after Kelly's police force in Chicago had indulged in their bloody orgy, Kelly came to Washington looking for political patronage. That patronage was forthcoming and Kelly must believe that the killing of the strikers is no liability in partisan politics.

"Meanwhile, the steel puppet Davey is still governor of Ohio, but not for long I think—not for long. The people of Ohio may be relied upon to mete out political justice to one who has betrayed his state, outraged the public conscience and besmirched the public honor.

"While the men of the steel industry were going through blood and gas in defense of their rights and their homes and their families, elsewhere on the far flung Committee for Industrial Organization front the hosts of labor were advancing and intelligent and permanent progress was being made. In scores of industries plant after plant and company after company were negotiating sensible working agreements. The men in the steel industry who sacrificed their all were not merely aiding their fellows at home but were adding strength to the cause of the comrades in all industry. Labor was marching toward the goal of industrial democracy and contributing constructively toward a more rational arrangement of our domestic economy.

"Labor does not seek industrial strife. It wants peace, but a peace with justice. In the long struggle for labor's rights it has been patient and forbearing. Sabotage and destructive syndicalism have had no part in the American labor movement. Workers have kept faith in American institutions. Most of the conflicts which have occurred have been when labor's right to live has been challenged and denied. If there is to be peace in our industrial life let the employer recognize his obligation to his employees—at least to the degree set forth in existing statutes. Ordinary problems affecting wages, hours, and working conditions, in most instances, will quickly respond to negotiation in the council room.

"The United States Chamber of Commerce, the National Association of Manufacturers, and similar groups representing industry and financial interests, are rendering a disservice to the American people in their attempts to frustrate the organization of labor and in their refusal to accept collective bargaining as one of our economic institutions. These groups are encouraging a systematic organization of vigilante groups to fight unionization under the sham pretext of local interests. They equip these vigilantes with tin hats, wooden clubs, gas masks and lethal weapons and train them in the arts of brutality and oppression. They bring in snoops, finks, hatchet gangs and Chowderhead Cohens to infest their plants and disturb the communities. Fascist organizations have been launched and financed under the shabby pretext that the Committee for Industrial Organization movement is communistic. The real breeders of discontent and alien doctrines of government and philosophies subversive of good citizenship are such as these who take the law into their own hands. No tin hat brigade of goose stepping vigilantes or bibble-babbling mob of blackguarding and corporative-paid scoundrels will prevent the onward march of labor, or divert its purpose to play its natural and rational part in the development of the economic, political and social life of our nation.

"Unionization, as opposed to communism, presupposes the relation of employment; it is based upon the wage system and it recognizes fully and unreservedly the institution of private property and the right to investment profit. It is upon the fuller development of collective bargaining, the wider expansion of the labor movement, the increased influence of labor in our national councils, that the perpetuity of our democratic institutions must largely depend. The organized workers of American, free in their industrial life, conscious partners in production, secure in their homes and enjoying a decent standard of living, will prove the finest bulwark against the intrusion of alien doctrines of government.

"Do those who have hatched this foolish cry of communism in the Committee for Industrial Organization fear the increased influence of labor in our democracy? Do they fear its influence will be cast on the side of shorter hours, a better system of distributed employment, better homes for the underprivileged, social security for the aged, a fairer distribution of the national income?

"Certainly the workers that are being organized want a voice in the determination of these objectives of social justice.

"Certainly labor wants a fairer share in the national income. Assuredly labor wants a larger participation in increased productive efficiency. Obviously the population is entitled to participate in the fruits of the genius of our men of achievement in the field of the material sciences. Labor has suffered just as our farm population has suffered from a viciously unequal distribution of the national income. In the exploitation of both classes of workers has been the source of panic and depression, and upon the economic welfare of both rests the best assurance of a sound and permanent prosperity.

"In this connection let me call attention to the propaganda which some of our industrialists are carrying on among the farmers. By pamphlets in the milk cans or attached to machinery and in countless other ways of direct and indirect approach, the farmers of the nation are being told that the increased price of farm machinery and farm supplies is due to the rising wage level brought about by the Committee for Industrial Organization. And yet it is the industrial millions of this country who constitute the substantial market for all agricultural products. The interests of the two groups are mutually dependent. It is when the payroll goes down that the farmer's realization is diminished, so that his loans become overdue at the bank and the arrival of the tax collector is awaited with fear. On the other hand, it is the prosperity of the farmer that quickens the tempo of manufacturing activities and brings buying power to the millions of urban and industrial workers. As we view the years that have passed this has always been true and it becomes increasingly imperative that the farm population and the millions of workers in industry must learn to combine the strength for the attainment of mutual and desirable objectives and at the same time learn to guard themselves against the sinister propaganda of those who would divide and exploit them.

"Under the banner of the Committee for Industrial Organization, American labor is on the march. Its objectives today are those it had in the beginning: to strive for the unionization of our unorganized millions of workers and for the acceptance of collective bargaining as a recognized American institution. It seeks peace with the industrial world. It seeks cooperation and mutuality of effort with the agricultural population. It would avoid strikes. It would have its rights determined under the law by the peaceful negotiations and contract relationships that are supposed to characterize American commercial life. Until an aroused public opinion demands that employers accept that rule, labor has no recourse by to surrender its rights or struggle for their realization with its own economic power.

"The objectives of this movement are not political in a partisan sense. Yet is true that a political party which seeks the support of labor and makes pledges of good faith to labor must, in equity, and good conscience, keep that faith and redeem those pledges. The spectacle of august and dignified members of congress, servants of the people and agents of the republic, skulking in hallways and closets, hiding their faces in a party caucus to prevent a quorum from acting upon a labor measure, is one that emphasizes the perfidy of politicians and blasts the confidence of labor's millions in politician's promises and statemen's vows.

"Labor next year cannot avoid the necessity of a political assay of the work and deeds of its so-called friends and its political beneficiaries. It must determine who are its friends in the arena of politics as elsewhere. It feels that its cause is just and that its friends should not view its struggle with neutral detachment or intone constant criticism of its activities. Those who chant their praises of democracy but who lose no chance to drive their knives into labor's defenseless back must feel the weight of labor's woe even as its open adversaries must ever feel the thrust of labor's power. Labor, like Israel, has many sorrows. Its women weep for their fallen and they lament for the future of the children of the race. It ill behooves one who has supped at labor's table and who has been sheltered in labor's house to curse with equal fervor and fine impartiality both labor and its adversaries when they become locked in deadly embrace.

"I repeat that labor seeks peace and guarantees its own loyalty, but the voice of labor, insistent upon its rights, should not be annoying to the ears of justice or offensive to the conscience of the American people."[1]

The Struggle of Labor in America

"I address you in the name of the Committee for Industrial Organization, and the millions of Americans who are looking to labor to give them leadership and hope for the future. Almost three-fourths of the people of America are represented by workers gainfully employed. The future of organized labor, therefore, is in broad sense the future of America.

"American labor is today in the throes of the greatest struggle in the history of the United States. The workers of America are becoming articulate. They are demanding their just share in the government of their nation. They desire to benefit not only themselves, but all other citizens of this country, by the establishment and preservation of a true democracy in their own country. The final determination of our endeavors lies many years in the future, but it is our firm belief that upon our success of our failure will depend not only the future of this nation, but in some degree the future of endangered civilization.

"The fabric of culture which has been built up by mankind through enduring centuries of painful toil and sacrifice is menaced today as never before. America is one of the group of remaining great democracies. If the citizens of the United States can pass on to their children the ideal and the practice of human freedom, we shall be not only fulfilling our duty as inheritors of the hard-won traditions of ages, but we shall be accomplishing our responsibilities to the future of our race.

"If the democracy of the United States is to survive and if its government is to be a true expression of the ideas of its citizens, there are abuses which must be corrected. Hundreds of thousands of the people of this nation have for years on years been exploited, oppressed, and denied the exercise of those rights guaranteed to them under the constitution. They have lived at the mercy of economic vicissitudes. They have been little more than industrial serfs. They have been unable to raise their wages, to ameliorate their working conditions to assure themselves of economic security. Millions have suffered the ills that accompany partial starvation in America, and the children of these men have been unable to improve their lot. It is not the purpose of the workers of the United States that these things should continue.

1. Robert William Deimel, *The Public Relations Activity of the United Mine Workers of America Between 1946 and 1950; A Contribution to the History of Public Relations* (1971), pp. 204-8.

"We do not intend that our people shall trudge aimlessly from street to street seeking only that work which they will never find because our economic leaders have neglected to provide it for them . We do not intend that our children shall starve in the midst of plenty. We do intend to take an active part in the government of our nation, so as to insure to the average citizen an improved participation in its economic and political bounties. For these purposes the workers of the United States have elected to form unions to which they flock by the hundreds of thousands. The labor movement in America has been reborn. Socially, economically, politically, the workers of America are becoming articulate. They must and will participate in the determination of their destiny.

"The Committee for Industrial Organization was formed by the heads of eight national and international unions in November, 1935. Its avowed purpose was the encouragement of workers in the mass-production industries to organize. Its success may be judged by the fact that its affiliates now comprise thirty-five thousand local industrial unions not affiliated to any national organization, and a number of state and local industrial union councils.

"The membership of the Committee for Industrial Organization now closely approximates four million.

"It is not my desire to dwell upon the division in the ranks of American labor. The members of the Committee for Industrial Organization at its inception desired only to work within the framework of the existing federation for the benefit of the unorganized workers in the United States. They were not permitted to do so. Our organizations were suspended from membership. After this action, we determined to extend our educational activities by independent organization of workers in the name of the Committee for Industrial Organization. The results of our efforts are apparent. In two years we have attained a membership of four million.

"It has been urged that labor should make peace. We are willing to do so. We believe that the craft unions have a definite place in the structure of our labor movement. We will not, however, submit to the destruction or to the dismemberment of our constituent industrial unions as the price of peace. The workers of the United States want to be organized. We purpose that they shall be organized. If this might be done under the aegis of a united labor movement, the members of the Committee for Industrial Organization would be more than jubilant, but since it cannot be so, we shall not submit ourselves and the welfare of the yet unorganized millions to the treacherous mercies of those false and feeble leaders who have spurned our honest overtures.

"The character and the success of the work of the Committee for Industrial Organization may be measured by the virulence of its enemies. We have been criticized by communists and fascists, by the National Asssociation of Manufacturers, by elements of the public press, and by reactionary public representatives.

"The Committee for Industrial Organization has been called

dangerously radical and dangerously conservative. The police forces of corrupt cities and the soldiery of some states have been marshaled against our people in their efforts to obtain for themselves increased wages, shorter working hours, better working conditions and a recognition of their right to bargain collectively. To be brief, we have been belabored by the abuse of all those who do not desire that the workers of America should improve their economic condition and by all those who do not wish these workers to proceed in an orderly fashion toward the attainment of such objectives.

"Our work will speak for itself. The activities of the Committee for Industrial Organization have raised the wages of American workers by billions of dollars. We have shortened their working hours and we have immeasurably improved their working conditions. We have provided for the adjustment of their grievances in their places of work. In many communities where workers have even been forbidden to exercise their right of political franchise, the oppressive political organizations of industry have been completely destroyed. Not less in importance to these chronicled accomplishments, is the fact that labor has become organized and articulate. It is henceforth capable of assuming a major degree of responsibility and leadership in the rehabilitation of economic America.

"Thirteen million Americans are now unemployed. Their numbers are steadily increasing, as the nation drifts with terrifying and deadly sureness to the never-never realm of financial bankruptcy, economic collapse and human tragedy. This appallingly true, despite the fact that government has dipped into the public purse to make possible the granting of huge subsidies to industry, agriculture, banking and finance.

"Since 1933 these grants, directly and indirectly, have amounted to the enormous total of twenty-two billion dollars. Our national internal economy has attained the amazing condition where it appears that practically all of our major enterprises are unable to exist or function on their own resources. Our banking structure is supported by federal funds. Our great shipping industry since 1917 has squandered nearly three billions of dollars derived during that period from direct and indirect government aid. In addition, it has squandered its own revenue and its capital assets, and a recent report of a government agency startingly reveals its financial prostration.

"Agriculture is the recipient of direct government aid, and the financial structure of our railway system is such as to require constant increases in freight and passenger tariffs to continue them in operation, all of which is added to commodity prices which inevitably increases the burden borne by the population. Countless business units in sundry industries are the recipients of huge government loans given under circumstances where the borrowing corporations are unable to qualify for loans from our banking institutions, which are themselves overflowing with money upon which they cannot pay depositors interest.

"Since 1933, approximately seven billions of dollars of government money has been paid to the unemployed in the form of direct or work relief. Most of this vast sum has trickled through the tills of the nation's shopkeepers and thus has become a subsidy to our merchants and professional men.

"America is moving in economic reverse. Our consumer goods industries began to slow down in June, 1937, and by October of the same year our heavy industries began to feel the icy hand of the depression. Since that time, the drift has been constant—shutdowns in industry being the order of the day—while the number of human beings thrown out of employment was augmented day by day. In the months that have ensued, neither industry nor government has come forth with constructive proposals, designed to meet the problems of the depression. The federal congress, lacking adequate or competent leadership, in continuous session for months past has failed to devise or enact a single statute that would cause a glimmer of hope to penetrate the minds of millions of despairing Americans.

"Meantime, cavilling and confusion prevails, and our statesmen and those carrying the responsibilities of the nation's manifold enterprises, are reviling each other with an anger and bitterness that defiles, sears and destroys. Meantime, the population suffers and a creeping paralysis progressively impairs its functions.

"What is to be done? Reason calls for a change. More rational policies are indicated. America is menaced, not by a foreign foe that would storm its battlements, but by the more fearful enemy of domestic strife and savagery.

"It is time for Americans to cooperate. It is time for Americans to recognize each other's right of individual existence. It is time for capital to recognize labor's right to live and participate in the increased efficiency of industry and the bounties of our national resources. It is time for labor to recognize the right of capital to have a reasonable return upon its investment. It is time for statesmen to recognize their nation's peril and to decide to cooperate with labor and industry, to rationalize the nation's processes and alleviate a nation's distress. Labor is willing to cooperate—now. Let the leaders of the nation's business step forward. Let the statesmen of the nation do the same. Let the council of reason and mutual toleration be convened. American leadership can accomplish this task, and in so doing will preserve its governmental structure and its democratic institutions.

"My compliments and good wishes to the people of the British Empire. Good night."[2]

2. Robert William Deimel, *The Public Relations Activity of the United Mine Workers of America Between 1946 and 1950; A Contribution to the History of Public Relations* (1971), pp. 209-212.

John L. Lewis Urges Congress to Furnish 3,500,000 Jobs

"Thirteen to fourteen million workers are now unemployed in this nation. That means unemployed who are able and willing to work. There are in addition countless millions who are employed only one, two, or three days a week.

"Unemployment of thirteen million means that there are at least five million who can be considered to be without resources of any kind, facing all the ravages of destitution. It is modest, indeed, that labor asks only three-and-a-half million jobs to care for thirteen million unemployed.

"In the spring of 1935 the federal government assumed the obligation to give jobs to the needy unemployed who were able to work. Since that time the administration has reiterated again and again the conviction that the Federal government must carry this responsibility. It has failed to do so. In establishing this policy the federal government threw back upon the states and counties total responsibility for those unemployed and destitute who could not work. But it has done more than that. It has thrown back upon the state and local governments, a burden that has now reached two million families. Probably at least sixty percent of the heads of these two million families are employable. Thus state and local governments find themselves being driven to the point of bankruptcy by a burden which the federal government has promised to assume. If it did assume such responsibility now more than a million workers would go on WPA payrolls. That would raise the required number of jobs beyond the modest three-and-a-half million which labor is asking. Even this statement does not picture the real need.

"The difference between the number of persons indicated as needing employment by calculating from the general employment figures and the number of persons on the WPA plus persons awaiting assignment is explained by the inadequacy of the relief machinery in the states. States are not able or will not put upon their relief rolls all those who need WPA jobs. This is due to limited personnel in relief offices, to certain local regulations and exceedingly inadequate funds for local relief. There is in addition a natural resistance on the part of self-respecting workers to face the heartbreaking and degrading process of getting on the relief rolls even though they be in dire need.

"Upon the basis of these facts labor has made its request to congress for provision of at least three-and-a-half million jobs for the unemployed on WPA. This is a minimum program. Any program that is expected to meet the basic condition that every unemployed person who is out of work through no fault of his own should have a job, would have to be nearly twice as great as the one labor asks.

"Nearly two months after the labor had announced the need for three-and-a-half million jobs the President set forth to congress his program. It is the belief of labor that this program substantially meets the request we have made. We are wholeheartedly in support of it. We believe that in order to make this program effective the executive should be given wide latitude in its expenditure. It is not possible at this time to forecast the nature of the emergencies which may arise in the next year. The people of our nation should not be made to suffer because someone may have guessed wrong. Labor believes that this program can be so operated, if proper authorization is given, that an average of three-and-a-half million jobs can be provided between the operations of the WPA and PWA for the next seven months which are covered by the WPA appropriation proposed in the President's Message.

"On two counts it is important that these three-and-a-half million jobs be provided immediately.

"First, there is terrible need among our people, and on all grounds of humanity and decency these human beings must be provided for.

"The second important purpose of the program is to provide and stimulate purchasing power and recovery. The longer that stimulus is delayed, so much longer do we drift not toward recovery but toward economic disaster. Therefore we cannot afford to wait for three or six months for the three-and-a-half million jobs to be provided. It may be too late then. Now is the time for action.

"Labor is asking that this fund be so appropriated that it can be operated under four main principles.

"The first principle is that all workers should be employed in work suited to their needs and skills. One of the fundamental purposes of the works program has been to maintain in our people their skills, their morale and their normal life. The program means conservation of human resources. If projects are set up upon which skills are destroyed, which require that the workers family life be broken up, this great purpose is not accomplished. If a man who has worked all his life as a clerk with his hand trained to pen and pencil, it is a harmful thing that he should be put to digging ditches. If a man is required to leave his family over long periods of time so that he may work on a project, that, too, is harmful.

"A careful study is needed to assure that there shall be projects to fulfill the needs and to conserve the skills of the unemployed workers. This implies foresight and planning. Useful projects cannot be created overnight. They must be ready when the emergency arises.

"The second principle upon which labor makes its demand is that projects should be socially necessary and productive. This means housing and slum clearance, flood control, the building of schools and hospitals, health measures, cultural and white collar projects, and vocational training work. The vast funds which our government spends upon the works program should not be wasted away on projects which fail to add to or conserve the material and human capital of our nation. That this purpose may be accomplished requires planning and foresight.

"Too long now our works programs have been created at a moment's notice. It is high time that provision is made for a foresighted planning of projects which will be not only useful, socially necessary and productive, but will provide work suited to the skills of those who become unemployed. It is time that there was established within the government a planning agency to determine where projects are needed and have ready at the time of economic emergency full plans so that work might be started immediately.

"The third principle that labor asks is that jobs should be given those who need jobs without requiring honest, decent unemployed workers to degrade themselves as paupers.

"I can conceive of no bitterer experience for a man than that of being as a pauper when he wants work and is willing to work. That tragedy is being enacted daily in our nation. If we wish to maintain that free American spirit of which we speak so proudly, then this brutal crushing of men's spirits must not be allowed to go on. Labor says give jobs to those who need them without making men suffer the degradation of being pauperized. Those who are most needy should be given first consideration. Those unemployed who are not on relief should not be forced to drag out their meager possessions to the last bitter end before they can receive help. As long as every unemployed person must be certified for relief before he can be sent to a WPA job this will happen. The immediate way to prevent this evil is to have the WPA make certifications directly.

"Another need arises in this connection. Millions of our working men and women now only have one, two or three days of work a week. This work in most cases provides income far below the standard necessary to maintain civilized life at a minimum standard of health and decency. Yet these men and women, though their income is less week to week and month to month than they get on the meager payments of the WPA are denied the right to get supplementary work on WPA. Nor can they give up their jobs and go on WPA. Labor asks that WPA provide supplementary work for parttime workers to bring their income up to an American standard.

"The fourth principle that labor asks is that funds be so appropriated that government will be able to meet current needs wherever and whenever such needs arise. The government should be equipped to meet emergencies.

"When three million men and women are added to the long list of those unemployed, as was the case in the last four months of the past year, the government should be in a position to meet the emergency unhampered by legalisms and red tape.

"At the present time also the financial affairs of many communities have reached such a pass that they are unable to provide the standard contributions for the establishment of WPA projects. The government cannot allow the workers in these communities to suffer. It must be prepared to enter such communities and establish its own projects even if adequate local contributions cannot be provided.

"One of the proudest boasts of our nation has always been the American standard of living. For those who are on WPA that American standard of living means an average of less than $52 a month income. That is an average of approximately $620 a year. In the southern wage region workers on WPA receive about half that amount. Unskilled workers receive as little as $21 a month and 15 cents an hour. Most of the men on these projects have families to support. Government has the obligation of creating decent standards. It now pays at levels below those of sweatshops, below those universally accepted as absolute minimums for American citizens.

"I cannot express in strong enough terms how inadequate these levels of income are to provide livelihood for American families. Such a condition can only bring a deep sense of shame to responsible citizens.

"Labor believes that congress is now confronted with a decision of profound importance. We are a nation rapidly slipping toward economic chaos. Up and down the land the mills and factories of the greatest industrial nation of the world lie idle while millions of our people wait for the right to work and to consume the products of their work.

"The President has enunciated a program of action to stir the nation from the economic doldrums. That program is closely in consonance with the measures which labor has been asking. I see no signs on the horizon of the nation which would indicate a natural course of recovery if congress chooses to remain inert.

"I solemnly say to you that this is the most serious decision which congress faces. If congress chooses to act in accordance with the President's program, our nation can be again turned toward economic vigor. If congress fails to act we will continue to drift in the direction of economic chaos. In that direction lies the greatest threat to democracy, the idleness and misery of our people. The life of any form of society depends upon its ability to provide security and happiness for its people. Fascism and anti-democracy feed upon hunger, hopelessness and desperation. American government must be prepared to act immediately and effectively in the interests of its citizens if we expect the traditional forms of American government and economy to survive.

"The progressive forces, labor and its friends, look upon this program as a test of their representatives at the nation's capitol. The passage of this program will be a vindication of their faith."[3]

Great Convention of Congress of Industrial Organizations

"President Fagan, His Honor, Mayor Scully of Pittsburgh, Father Rice, Rev. Simon, delegates to the first constitutional convention of the C. I. O., and the United Mine Workers' Band of Bridgeville, Pennsylvania, ladies and gentlemen—I profoundly appreciate the opportunity of opening this convention with greetings through President

3. Robert William Deimel, *The Public Relations Activity of the United Mine Workers of America Between 1946 and 1950; A Contribution to the History of Public Relations* (1971), pp. 213-16.

Fagan of organized labor in this great industrial section, with the greetings of his Honor, the Mayor of Pittsburgh, extended in behalf of the people of this great municipality, with the greetings of the Christian churches, represented by the eminent clergymen who are present this morning, and the acclaim of our own people. Why these greetings? Why this interest? Why this enthusiasm? Why this acclaim? Because there has been born in America a new, modern labor movement dedicated to the proposition that all who labor are entitled to equality of opportunity, the right to organize, the right to participate in the bounties and the blessings of this country and our government, the right to aspire to an equality of position and the right to express views, objectives and rights on a parity with any other citizen, whatever may be his place, his condition of servitude, or the degree of world's goods which he may possess.

"So that is the greeting of the Committee for Industrial Organization, assembled here and about to formalize its own internal affairs and make permanent its form of organization.

"It is perhaps an interesting coincidence that 57 years ago almost to the day the great Gompers founded in this city the labor movement of his generation. That labor movement served that generation in a period where the skills in American trade and industry were the skills of handicraft and not the skills of the machine age and of mass production. But time moves on and the old order changes, and as the changes became obvious it was more and more apparent that the labor movement and the type of organization or rendering service to the teeming millions who labor in American industries in this generation of our life.

"Perhaps it will be illustrative to say that in 54 years of existence and advocation and administration of its affairs the American Federation of Labor failed to bring to the hundreds of thousands of workers in the industries in the Pittsburgh area the blessings of collective bargaining, and during all those years the Pittsburgh industrial district was the citidel of non-unionism in America, the citidel of labor exploitation, and the recognized fortress of those financial interests and industrial interests in America who preferred to exploit and debase and degrade labor rather than recognize its existence or concede its right to fair treatment. The old order changes, and what the American Federation of Labor could not do in 54 years of agitation the Committee for Industrial Organization has done in less than three years. Does that meany anything?

"Ask the workers in these great steel and other industrial plants within 100 miles of downtown Pittsburgh. They will tell you whether or not it means anything and they will answer the question why they joined the CIO, they will tell you that they were not receiving the aid and succor from the American labor movement which they should have received.

"The Pittsburgh area today is the most completely organized city of any city or any area in industrial America. Whether you come into

Pittsburgh through the Ohio Valley, the Monongahela Valley, or the gateway of the Allegheny, you pass along miles of great industrial plants, and when you pass one of these great industrial plants in coming through any of those gateways, you pass a plant where the CIO has established collective bargaining, and where the employees there are members of the CIO.

"To you, Philip Murray, Vice-President of the United Mine Workers of America, Chairman of the Steel Workers Organizing Committee, my compliments, sir, for this super-human task that you have accomplished in the area where you live, where the headquarters of your great organization exist. You have rendered a service not alone to the people of Pittsburgh, but to the workers in these great industries, you have rendered an outstanding and superb service to the labor movement of America, to our country and to our flag.

"In the light of that record which I have merely portrayed as indicating the events that have transpired in this area, and which may be duplicated ad infinitum in other areas, in the light of that record, why should the CIO as a movement be criticized, opposed, slandered and vilified, denounced from the street corners by its adversaries, and constantly opposed in high places when it offers to the community, to the state, and to the nation a program of rational procedure and orderly conduct, a program of working out in a peaceful way the problems that encumber the relations of labor and industry and finance in this country.

In this steel industry to which I may aptly refer because of our presence here today, Mr. Murray and the Steel Workers organization have contracts with more than 540 corporate entities engaged in the manufacture and the fabrication of steel, and in every one of those plants in every one of those corporations without exception, since the negotiation of those agreements, peaceful relations have obtained and mutual satisfaction prevails between management and labor—something of a record of an industry that for a lifetime, through oppressive measures, prevented the organization of its employees and denied to them the right to join the union of their choice.

"But why should there be opposition and criticism of a movement that stands for orderly procedure and for a rational working out of the problems of modern industrial relationships, of an organization that is dedicated to the proposition of maintaining and supporting our democratic form of government, of an organization that is dedicated to the proposition of the right of investors to have a profit on their investment, and an organization that maintains the right to the freedom of contract relations between citizens of our republic, that is willing to lend its strength and its resources and its young men at any time to support and maintain that form of government, asking only in return that the safeguards of the constitution and the Bill of Rights be extended to cover the most lowly, humble worker, as a right, as a privilege for an American.

"There are troublous times in the world of affairs. Great and sinister forces are moving throughout the world, and he is optimistic indeed who believes that those forces will not affect Americans and will not have their impact and repercussions upon the peoples of the Western Hemisphere. Democracy is on trial in the world and in the United States. We want to preserve democracy. We cannot preserve democracy here in our country if we encourage as a people the overwhelming tidal wave of criticism, slander and abuse for an American institution like the CIO that stands for the protection of the privileges of all Americans whether they be gentiles or Jews or of any creed or religion, or any school of thought that maintains its self respect for our institution.

"We stand appalled today at what we witness in Europe. Whose heart can fail to become anguished as he reads in the daily press of the terrible abuses and atrocities and indignities and brutalities that are now being inflicted by the German government and some of the German people on the Jews of that nation? One of the most appalling events in history, shameful indeed to our concept of the ethics of our modern civilization which we boast, harking back to the practices of the medieval ages, the torture and debasement of a great race of people who only ask the right to live.

"Our Declaration of Independence says that we hold all men to be created equal. That means regardless of his creed, color, his race or his nationality. We foregather under that flag and we proclaim that creed, but that principle is being made a mock in the Germany of today. In Germany the labor movement was first wiped out and its leaders were harried and sent to concentration camps, and now in progressive fury and increasing brutality the German government is found inflicting these pogroms on the Jewish race.

"I say to my fellow countrymen, and I say it to the rich and influential and wealthy gentiles of America as I say it to the rich and influential Jews of America, you cannot strike down in this country through the use of your influence, great as it may be, a powerful movement of the workers of this country under the banner of the CIO, who stand for equality of protection to any group, any minority, any religion that exists here in our country.

"The United States of America is under increasing pressure in the realm of foreign affairs. The United States of America may one of these days face a great external crisis. When this mad, blood-thirsty, wolf of the German government inflicts its will upon the defenseless people of Germany, of Austria and of Czechoslovakia, and incites individuals in other countries to perpetrate the same atrocities in Europe, then it is possible that we will have to meet the German dictator as he tries to extend his domain into the realm of the Western Hemisphere. If that day comes, who is going to sustain the United States of America? Who is going to man its industries? Who is going to send its young men to military ranks to engage in war? Labor—labor! Who is going to protect the institutions of this country, those that are meritorious? Labor! Who is going to protect the titles to property and great wealth down through the generations in America? Labor!

"Who is going to do the suffering and the dying in the future but the sons and the daughters of the workers of this country? The workers of this country will never make anything out of war, they merely work and sweat and fight and die. Some one else takes the profits. Who took the profits in the last war? Not labor. And if war comes the United States needs the co-operation of the millions of workers that are members of the CIO.

"In consideration of all of these things, in consideration of the fact that we are Americans, and that we believe in the principles of our government, that we are willing to fight at any time to maintain that flag, we are going to ask from those who are the beneficiaries of that service and that attitude and that policy and that loyalty, we are going to ask proper treatment ourselves,—proper treatment ourselves! And I have every confidence that our government and our state department will make emphatic representations to the German government, protesting the actions of that government in permitting these atrocities to be inflicted on the Jewish people. I say to the government of the United States if, as and when it takes that action, the twenty million members of the CIO and their dependents will support the government and uphold its hands.

"The old order changes. Neither opposition nor misrepresentation is going to destroy the existence of the CIO as a living entity, and the people in high places in America who are now using their influence, through great newspapers and publications, who have opened the sluice gates of adverse propaganda against the CIO in America will awaken to the fact that their efforts are future and that the day may come when they, individually, will be rushing to the CIO and begging for its protection for their special privileges and their wealth.

"I express to you all my appreciation of your courteous attention while I have made these few remarks. I express with you our hope that from the deliberations of this convention may come renewed inspiration and greater confidence for the millions of workers of this country who are looking to the CIO to give them aid and assistance.

"We must not forget that we have 12 million more or less unemployed people in America who have a right to work but who have not been given work; that there are dependent upon those 12,000,000 many more millions who are underprivileged, ill-provided for and who are asking for a participation and are looking to this convention to devise policies, to state objectives, to lay out procedure that will cause them to have hope for the future.

"I am sure that every man and woman here recognizes the great weight of responsibility upon them in their representative capacities, and that each will contribute toward the successful foundation and the final completion of this great, new, modern labor movement of the CIO."[4]

4. Robert William Deimel, *The Public Relations Activity of the United Mine Workers of America Between 1946 and 1950; A Contribution to the History of Public Relations* (1971), pp. 217-20.

President Lewis Stirs Auto Workers in Address on Peacetime Draft

"I doubt that some of the proponents of universal peacetime conscription really understand what they purpose, or the implications or the potentialities of their own proposals. The chief of staff of the United States army nonchalantly comes before a committee of Congress and proposes a standing army of 2,000,000 armed men, quartered on the citizens of the United States. I wonder if he understands the social and the political implications of such a suggestion! I wonder if they understand the economic implications and potentialities of such a program!

"Build up a defensive army of that magnitude and train our millions of young men in the arts of war at the public expense, and as sure as the sun will rise tomorrow, by that act our Congress is planting the seeds of destruction of democracy and is paving the way for the rise of a new imperialistic nation within the confines of the United States of America.

"Build up, if you please, a gigantic military instrumentality of that character and quartered upon the people of this nation under a Roosevelt or under any other President, call it a defensive mechanism, if you please, but sooner or later will come a chief executive of the United States, a man on horseback who will believe this instrumentality is not a defensive instrumentality at all, but an offensive instrumentality that will carry out his imperialistic dreams and conceptions.

"And after all, our forebears who came to the shores of America from the nations of Europe and elsewhere came here to escape militarism. They came here because the citizens of many of the countries of Europe had to carry from one to three soldiers on their backs, and with the excessive taxation that followed the necessity of supporting such a machine, with the inevitable limitation of civil liberties with the constant enactment of drastic laws to enforce the growing power of the military caste, they found it necessary to leave, to flee and to escape to the land of the free and the home of the brave.

"What is our Congress and our country coming to when men in Congress propose as an amendment to the conscription act that anyone who would express an opinion in opposition to conscription of the administration of the law be liable to a fine of $10,000 and five years in prison! I wonder if, in our zeal to protect our country from the shadow of a Hitler, we are not foolishly taking on some of Hitler's own methods to build up that protective situation which, in the end, may circumscribe our individual liberties and eliminate our collective civil liberties, and put a burden of taxation on the people that will be oppressive and that will create social confusion and unrest to a point where we are in fact bringing about the condition from which we have been vainly seeking to escape.

"It is proposed that the young men be drafted into the army and that they be paid the army wage of $21.00 per month. What young man who has a father or mother or a sister or other kinsmen or relatives can idly contemplate entering the army for one year or three years and abandoning his family and his obligation for the sum of $21.00 per month?

"The Congress of Industrial Organizations in a letter yesterday to the chairman of the two military committees of the House and Senate, suggested as a practical proposition that the defense mechanism be augmented by voluntary enlistment for one year at an increased wage over and above the $21.00, and if that suggestion is accepted by the Congress, I am confident that our defensive mechanism will be well manned and the country will be saved many millions of dollars of expenditure, the country will be spared the grave social and economic and political repercussions that would follow an ill-advised course of that character in this matter.

"The Congress of Industrial Organizations wants the support of the United Automobile Workers of America and its great membership on this question. It affects you, and you, and me and every other American. Let your Congressman know, let the people of this country know that you stand for the essentials of democracy, and that the young men of this nation stand for participation in the freedom and in the liberties and in the privileges that exist under this flag."[5]

Not Guilty

"For a year and more, I have been branded Public Enemy No. 1 and the 600,000 members of the United Mine Workers have been stigmatized as malcontents who put mean greeds above the welfare of their country. Political malignity, springing from a determination to destroy all who cannot be controlled, has conducted an organized campaign to stir a fury of rage against us, both at home and on the firing lines abroad. No other labor executive or union has been subjected to any such bitter, cruel and sustained attack.

"Countless strikes, many for reasons shocking in their essential trivality, have disrupted and are disrupting the nation's war effort, but neither leaders or strikers have been named and pilloried. No clarion from the White House starts the hue and cry against them by accusations of disloyalty and sabotage. Miners alone, for pressing just demands, have been singled out for hate and obloquy.

"Fortunately, there is a record that gives the lie to every charge leveled against our patriotism. In 1943, when our work stoppages were alleged to have cut production down to the danger point, we mined 9,000,000 more tons of bituminous coal than in 1942, and with 25,000 fewer men. For the first four months of 1944, we produced

5. Robert William Deimel, *The Public Relations Activity of the United Mine Workers of America Between 1946 and 1950; A Contribution to the History of Public Relations* (1971), pp. 221-22.

10,244,000 tons more than in the comparable period in 1943 and with 45,000 less men.

"The same increases obtained with respect to anthracite, despite the loss of 18,000 workers. Aged miners, many in their seventies, came out of retirement to take the places left vacant by the drafting of 86,000 sons now fighting on every battle front.

Miners Victims of Inspired Propaganda

"Does it mean nothing that this great underground army, tricked by bureaucrats and damned as traitors by political enemies, stayed on the job, toiling as never before? And this while other defense industries were paying two and three times more than miners' wages to inexperienced boys and girls! Is it without significance that after eight months of trickery and strategems, our basic contentions were upheld in detail by court decisions and official findings? Deliberately, and choosing my words with care, I charge that our differences with the operators could have been settled quickly and amicably but for the persistent interference of certain government agencies.

"Inspired propaganda is responsible for a general belief that the United Mine Workers took advantage of the national emergency to make exorbitant wage demands—blackmail, to put it plainly. Nothing is further from the truth. On March 10, 1943, twenty-one days before the expiration of our contract, we sat down with the operators to argue out an agreement that would cover another two-year period.

"The United Mine Workers, I might point out, is the only labor organization that enters into contracts covering any such length of time.

"At the outset of the conference, we asked for a wage adjustment that would lift the daily rate from $7 to $9. As an answer to the cry of blackmail, let me cite some of the reasons behind our request.

"Under our contract, the miners worked a five-day week. Not that this was our choice. Always, and particularly at the outbreak of war, we had urged a longer working time. Assuming employment for 260 days at $7 a day that would be an annual wage of $1,820. But due to accidents, shutdowns, and repairs, the average for the year was 231 days, and the average annual wage $1,617. Does that seem a princely sum for a man with a family?

"Even this amount, however, was not his "take home" pay. Tonnage men were required to pay for their tools, blasting powder, fuses, blacksmithing, the upkeep of pit lights, the rental of their electric caps, and the salary of a checker. In addition, they had to buy special safety clothing, not the least item being three or four pair of expensive boots during the year. By reason of these deductions, the net annual average wage of a coal miner ranged from $1,000 to $1,700.

"These wages, poor enough in peacetime, fell below a decent subsistence level in wartime. Mines, unhappily, are not located in or near populous centers, but in remote places where competition does not operate to keep down prices. Of the 4,200 "company stores" in the United States, 3,500 are in the coal regions and all are a source of rich profit to the companies. By 1943, prices in these stores had increased by as much as 100 percent entailing actual privation. At the time of my appearance before the Truman Committee, Senator Ball accused me of demagoguery when I stated that miners were hungry. As my own testimony may be attacked as prejudiced, I will call to the witness stand Mrs. Agnes Meyer, the distingushed wife of Eugene Meyer, owner of the Washington Post, a gentleman open in his denunciation of the United Mine Workers. It is hardly conceivable that she can be accused of partisanship.

Lunch Boxes Were Empty

"Mrs. Meyer, after an inspection tour through the Pennsylvania coal districts, published these findings: "When I looked into those pathetic lunch boxes, I was ashamed. One man had two ears of corn and two chunks of bread without butter. Every box had starches but very few proteins and not enough fats. . . . He (the miner) is actually worse off than he was in 1941, much worse off. Not only has the price of the miner's food gone up. The rubber shoes he has to buy are up from $3.60 to $5.90. The miner's heavy shoes are almost $2 more than they used to be. . . . They are supposed to do the dirtiest, most hazardous and most arduous job in industry on pay that does not compare with that in steel mills, construction work or white-collar jobs. As I have seen plenty of morons in the Southern shipyards, who are not worth a nickel of anybody's money, getting twice the pay of these trained men, I felt the deepest sympathy for the miners."

"Even this is not the whole story. In no other industry in the world does a man report for work, stand in line to get his tools and equipment, take his turn on the small cage to do down into the deep recesses of the mine, and then ride and walk long distances, all before his pay starts. The average distance from the pit mouth to the "working face" is often between four and five miles, but in many instances it runs as high as ten or eleven. As a result, miners often spent 52 hours a week underground, and were paid only for 35 hours.

"Moreover, this travel back and forth, morning and evening, is the hardest and most dangerous part of the day's work. First, the risk of derailments and sagging trolly wires, and then the drudging, stumbling walk from central unloading points to the working face. Up to the time of the Italian campaign, deaths and injuries in American coal mines exceeded all casualties in our military forces. In 1943 alone, they reached a total of 73,482 fatal and non-fatal accidents.

"These conditions had become intolerable by 1943, but we were bound by a contract, and it is the pride of the United Mine Workers that never once has the organization breached an agreement in the fifty-three years of its existence. It was to a new contract that we looked for relief, with full confidence that it would be granted, for the whole situation had changed in a very vital respect.

"In 1938, the Fair Labor Standards Act stated specifically that travel time in a mine should be regarded as working time. A United

States District Court had upheld the law in a case involving iron-ore miners, and in 1941 the Fifth Circuit Court of Appeals sustained the decision. I might add, parenthetically, that the Supreme Court of the United States, in May of this year ruled that portal-to-portal time constituted 'compensable work.'

"When, therefore, we faced the operators on March 10, the $9 day was not regarded as a wage increase by us, but as compensation commanded by law, and upheld by two judicial decisions. Even so, it was not put forward as a demand, but as a basis for bargaining. Instead of facing us as a body, however, willing to work out a wage agreement that would cover the nation, the Southern companies split off from those of the North, forcing us to run from one to the other.

"Worse still, it soon stood apparent that neither group had any thought of bargaining in good faith, both rejecting our proposals without a single counter-suggestion. Secretary Ickes, looking back over the record at a later date, stated flatly that the operators went into conference 'determined to block agreement.'

"By March 22, when no progress had been made, President Roosevelt asked that negotiations be continued beyond the time of contract expiration. We agreed at once to an extension until April 30, but the attitude of the operators remained unchanged. Their secret purpose was soon revealed, for on April 22, without a warning word, Madame Perkins certified the case to the War Labor Board. I said then, and I say now, that it was a brazen breach of faith, for our agreement to the thirty-day extension was based on the explicit understanding that negotiations would continue. But that was not our only reason for refusing to acknowledge the jurisdiction of the Board. Our other principal objections were these:

"An utter lack of confidence in the Board, proceeding from the deep conviction that it was a hodge-podge body whose decisions went by favor. Bitter experience had taught us that we were not among the elect. The fact that the Board was on record with an opinion that the $2 increase was not an adjustment, ordered by law and upheld by the courts, but a wage raise, consequently forbidden by the Little Steel Formula. What we were asked to do, therefore, was plead our case before a tribunal that had decided against us in advance. It was because of this knowledge that the operators sabotaged collective bargaining for forty-two days, plotting all the while to drag us before a packed court.

"The decision to ignore the summons of the Board, I may point out, as not my own fiat. It pleases politicians, press and radio to paint me as a dictator, but nothing was ever more false. The striking characteristics of coal miners are intense individualism and fierce independence. They are men who resent any departure from democratic procedure. The actions of the organization are determined by a Policy Committee of 225 men, drawn from every section of the country, each one of them a miner or a man who has been a miner. No high-priced corporation lawyer sits in our councils or acts as our mouthpiece.

"It was the Policy Committee, by unanimous vote, that rejected the jurisdiction of the War Labor Board and demanded the resumption of collective bargaining. The demand was not heeded, and when the thirty-day extension ended on April 30, the miners laid down their tools. This is our historic and automatic procedure. For fifty-three years, a contract has been our protection against rapacity and ill-faith. No contract, no work. No stoppage order ever goes out. If a new agreement has not been signed before the expiration of the old, the men quit. Nevertheless, we were not careless of the national interest, for we knew that there were 79,000,000 tons of bituminous coal in storage.

"On May 1, Secretary Ickes took over the coal mines in the name of the United States. At once I asked for a conference and on May 2 I sat in conference with him, in his capacity as Solid Fuels Administrator. He asked me for an immediate resumption of production, and my agreement to this was sustained by the Policy Committee the same day. Nevertheless, President Roosevelt went on the radio that very night and, in utter disregard of our action, appealed to the coal miners to get back on the job.

"On May 17 another extension of fifteen days was arranged with the Secretary, a total of thirty days during which miners produced as never before, hopeful that a contract would be negotiated. Let me say that there was not time that Mr. Ickes did not exhibit good faith, but at every turn his hands were tied by the petty vanity of the War Labor Board. At every intimation of action on his part, Chairman William H. Davis would rush to the White House and win an order from the President that reduced the Secretary to a 'dummy' status.

"On May 25, lo and behold, the W. L. B. came out with a 'decision' in our case, and one based, incredibly enough, on the findings of a panel of three, not one of whom had ever been in a coal mine. And what was this mouse that the mountain brought forth? The vacation compensation of coal miners was raised from $20 to $50, and the companies were ordered to assume the costs of safety devices, blacksmithing, lamps, explosives, etc.

"The vital matter of portal-to-portal pay, however, was dodged, the Board holding that it presented 'a mixed question of law and fact . . . which can and should be settled by collective bargaining.' In the same breath, however, forty-eight minutes was put forward as the total daily travel time of a miner, a figure gathered from a report made six years before by a college professor, and covering only a few mines. And this in face of admission by the operators themselves that the average travel time was one hour and twenty-five minutes!

"But did the Board order the operators to resume collective bargaining? Not by so much as a whisper! With contract negotiations suspended for a second time, the miners laid down their tools. Straightway the President commanded a return to work by June 7, and Secretary Ickes wired a request that I 'direct' such return. I answered that I had no power to 'direct,' but that I would be glad to recommend.

New Hope for Agreement

"At a stormy meeting, the Policy Committee voted a return to work at once, but on these two conditions: (1) that the government order the operators to obey the law of the land and enter into conference with us; (2) that the government retain control of the mines until a contract could be drawn and signed. Anything else would have meant rebellion, for our membership had grown increasingly bitter and distrustful.

"Grudgingly enough, the operators sat down with us on June 5, but before a mouth could be opened, the War Labor Board ordered an adjournment. And the reason? Because the miners were not yet back on the job! Not important, perhaps, although it lost two days, but it showed the animus of Chairman Davis and his strutting little men. Permitted to assemble on June 7, the operators presented their usual sullen, stubborn front, dominated by ex-Senator Edward Burke, the $25,000-a-year spokesman for the Southern Coal Producers Association.

"In mid-June, the first timid ray of light broke through our gloom. The Central Pennsylvania operators, tiring of the intransigeance of their associates, broke away from the combine and offered us a contract. By its terms, portal-to-portal pay was fixed at $1.25 a day, to which was added the increased vacation compensation authorized by the W. L. B., free tools, safety devices, etc., making a total of $1.50. Not the increase for which we had hoped or to which we were entitled, but the Policy Committee accepted, trustfully expectant that it would open the way to peace.

"And then, before the contract could be signed, 'hell broke loose,' to use the words of one unhappy operator. Railroads, steel companies and other principal customers of the Central Pennsylvania group, kept the wires hot with threats and, under this compulsion, the offer was withdrawn.

"Other evidences of ill-faith added to the anger and unrest of our membership. Some months before, a six-day week had been agreed upon, and to take care of the overtime expense for this sixth day, the O. P. A. had permitted an increase of from 10 to 23 cents a ton in the selling price of coal.

"Instead of making good on their bargain, however, the operators blithely pocketed the increase, and either refused to let the miners work the extra day, or else staggered employment so as to stay within the limits of the old five-day week.

"Now for a proof of the miners' faith that was not mentioned at the time by press or radio, and which has gone unnoticed to this day. Secretary Ickes, while admitting that portal-to-portal pay constituted a legal obligation under the law and court decisions, insisted that he was without power to make such payment. On June 20, therefore, the United Mine Workers voted to waive such payment from that date on until the end of government operation. A gift of millions, as it turned out!

"The days went by, each more profitless than the other, but on July 21 came a second break in the phalanx formation of the operators. The Illinois operators, frankly confessing a desire to settle portal-to-portal compensation by compromise rather than litigation, presented a contract much the same as that offered by the Central Pennsylvania group; 1$1.25 a day for travel time, plus the extra 25 cents permitted by the W. L. B. In addition, however, there was a pledge to work the six-day week. And once again the War Labor Board took upon itself the responsibility of continuing strife, forbidding signatures to the contract on the ground that it provided for wage increases in violation of the Little Steel Formula.

"July gave way to September and October, and still the operators sabotaged every approach to agreement. Many of them crept to our doors with expressionless regret and dissatisfaction, but Senator Burke and the Southern operators held them in line by assurances that the War Labor Board stood behind them, and would never surrender to John Lewis and his gang. On October 13 Secretary Ickes, doubtless sick of trying to administer under the burden of handcuffs and leg chains, turned back the mines to private operators.

Bureaucratic Meddling, Hairsplitting Legalisms

"Now for the capsheaf to bureaucratic meddling and sly malice. On October 26, two months after its rejection of the Illinois contract, the War Labor Board pontificated the reasons for it. There is no point in any discussion of its details, for I myself after repeated readings, am unable to make head or tail of its tortuous, hair-splitting legalism. About all that stood clear was the Board's opinion that the $1.50 offer represented a wage increase of 37½ cents a day, and could not possibly be allowed. 'The Board could approve a daily wage of $1.12 but no more.' By the same twisting and turning, anthracite miners were ordered to accept a 32.2 cents increase per day, less than half of what had been asked.

"On November 1, 530,000 bituminous and anthracite miners 'took a holiday' of their own accord. No single order issued from any official source, either nationally or locally. It was the unanimous protest of men who were tired of serving as guinea pigs for Washington's campus theorists, and sick of sabotage and double-crossing. More than that, it was the answer of the rank and file to the repeated charge that they were not behind their elected leaders, and that our demands were 'purely political,' put forward by me to embarrass and harass the Roosevelt Administration.

"Again the government took over the mines, but this time with a difference. The President's order authorized and directed the Secretary of the Interior, 'to offer to the duly constituted representatives of the workers' own choosing a contract or contracts governing the terms and conditions of employment for the period of the operation of the mines by the government.' Here, at long last, was what we had been fighting for! The chance to sit down at a table and work out an agreement by collective bargaining. It was what Secretary Ickes had urged in May, and what the War Labor Board had blocked out of vanity and venom.

"Two days of conference followed—frank, vigorous discussions, but with both sides searching for agreement—and on November 3, a contract was signed between the United States and the United Mine Workers of America. Well knowing that the War Labor Board, backed by the White House, would stand on the $1.12½ a day authorized in the Illinois decision, we accepted that dishonest figure, but upped it to $1.50 by consenting to take fifteen minutes away from lunch time and add it to 'productive time.'

"On November 5, the War Labor Board approved the contract, but in language so grudging that it opened the door for further contention by Senator Burke and the southern operators. As a consequence, this group persisted in recalcitrance and refused to join with other operators—representing 70 percent of coal production—when a contract was drawn up between the U. M. W. and the employers. This 70 percent, however, hailed the terms as fair, and expressed regret that the agreement had not been reached months before.

Slow-Moving Bureaucracy

"But did the War Labor Board take quick action on this contract? Although the contract was signed by both parties on December 17, it was not until May 20 of this year that approval was finally given. For five months the Board sat on the agreement with all the dull persistence of a hen trying to hatch a porcelain egg.

"With what result? The Southern group was encouraged to balk, boggle and muddle, and the operators as a whole were permitted to welsh on an acknowledged obligation. On October 26, Secretary Ickes, with the unreserved approval of the War Labor Board, had fixed on the payment of $40 to each miner as settlement in full for travel time between March 31, 1943, and June 20, the date when we waived payment by the government. Taking advantage of the W. L. B.'s procrastination, the operators withheld the payments, and at this writing are still retaining possession of $18,000,000 of our money.

"There, then, in brief, is the story of our "treason"—the fight of more than half a million Americans, working in the world's most dangerous and arduous industry, to win a daily wage that is above subsistence level!

"A fight carried on against political power and employer sabotage, aided by press and radio with their campaigns of misrepresentation! A fight to uphold collective bargaining as opposed to bureaucratic fiat! And tragically enough, a fight settled on terms that could have been had at the very beginning but for malign interferences that put the destruction of the United Mine Workers above the national interest.

"Today, under a contract far from meeting our ideas of fairness, we carry on in the darkness of the underground, faithful in observance of that contract down to the last detail, doubling production. Meanwhile, other industries are disrupted and bedeviled by walkouts and work stoppages in insolent disregard of contractual obligations. But they are not stoned and execrated as were the United Mine Workers."[6]

Lewis Warns That Legislative Shackles On Labor Mean Doom of Free Enterprise

(The following are excerpts from the text of President Lewis's statement before the U.S. Senate Labor and Public Welfare Committee on Friday, March 7, 1947.)

"I appear in opposition to the enactment into law by the Congress of regulatory, punitive, and restrictive laws dealing with labor unions and their memberships. We are the one great country in which economic freedom still exists. The first duty of the Congress and all of the citizens of our country is the conservation of this priceless heritage.

"We have just fought another great war to preserve democracy and freedom of action. We have just demonstrated to the world that democracy can function, even more efficiently than the absolute forms of government. Our productive record was the greatest miracle in the history of man. The sheer weight of our production, produced by free workers in America, certainly exceeded the production in any other nation, no matter what form of government it had.

"The United States entered this World War with certain laws guaranteeing and implementing the rights and privileges of labor unions to organize and to bargain collectively for the welfare of their members. Now, after the winning of this, the greatest war the world has ever known, it is now proposed that the liberties and rights of these workers who have performed such a service shall be limited, restricted, and annihilated.

"The kind of economic government that a nation is to have is a major decision which the people of such nation must make. The democratic way puts control over the enterprise in the hands of those essential to its operation, each group responsible for its own function. The bureaucratic or totalitarian way puts all groups under outside control. There can be no compromise between these two groups.

"The 80th Congress can strengthen our free enterprise system by its action in maintaining the economic freedom and liberties of the workers. The 80th Congress can also strike what might well be a fatal blow to this system by attempting to shackle the workers with restrictive and punitive laws. . . .

Federal Mediation Board and Cooling Off Period

"The proposal to set up a Federal Mediation Board with a high-priced personnel subject to political appointment is unnecessary and

6. Robert William Deimel, *The Public Relations Activity of the United Mine Workers of America Between 1946 and 1950; A Contribution to the History of Public Relations* (1971), pp. 223-30.

constitutes more government control by a bureaucratic agency. A conciliation service in the Department of Labor, properly strengthened, is sufficient for the needs of conciliation and mediation in labor disputes. Any independent board would simply be carrying over into peacetime the regulations and red tape that characterized the wartime agencies, and which the people are so emphatically insisting shall cease. Labor has had enough of these boards and bureaucratic meddling in its affairs.

"You are proposing to create by this bill another great governmental bureau that will start with hundreds and grow into thousands of civil servants and non-classified employees of the government at the public expense. This is highly inconsistent with the talk of consolidating government bureaus and doing away with unnecessary controls.

"As an example of what this board would probably grow into, take the War Labor Board. This Board started with a very few employees, and grew into a bureau that has 8,000 full-time employees and 12,000 panel members in the several states, consisting of lawyers, college professors, politicians and so-called economists. Some 12,000 employees of a little governmental board that did nothing to contribute to the economic well-being of America and that tried to fasten of its own act compulsory arbitration upon the worker of this country.

"Under the hypocritical claim of mediation and cooling-off periods, a board would be set up by this bill that would in truth and in fact be the first step toward imposing compulsory arbitration upon the millions of workers in this country, and would be the first thrust of the knife of absolutism into the heart of free America. . . .

Prohibitions and Restrictions Against Contracts for Health and Welfare Funds

"The problem of health and welfare of the workers in the country has become one of the major problems of the whole national economy. It is one of the vital problems that must be solved if we are to maintain the free enterprise system. The success or failure of a solution of problems of health and welfare of the workers by collective bargaining might well determine the extent to which the government may intervene in the field of health and medicine. It therefore baffles the understanding as to why anyone should propose to prohibit or to limit by law an attempt to provide for the health and welfare of the workers by mutual and satisfactory agreements between the parties.

"The United Mine Workers of America negotiated an agreement with Secretary of Interior Krug, acting as Coal Mines Administrator, on May 29, 1946, which contained a provision for the creation of a health and welfare fund. This fund is to be used for making payments to miners and their dependents and survivors with respect to (1) wage loss not otherwise compensated at all or adequately under the provisions of federal or state law and resulting from sickness (temporary disability), permanent disability, death or retirement; and (2) other related welfare purposes as determined by the trustees. . . .

Inadequate Hospital Facilities

"The mine workers' investigation disclosed further that in most districts miners participate in group hospitalization plans, but that the hospitals entering into these agreements are, for the most part, virtually inaccessible owing to their distance from the mine, ranging up to 160 miles; that they are understaffed; lacking in modern equipment and ambulance service. For the poor service provided them and their dependents by hospitalization insurance, miners pay an average of $1.95 monthly.

"The overwhelming majority of local unions in the Southern districts maintain burial funds, supported through monthly wage deductions. The burial check-off averages $1.29 per month per miner.

"In addition to what he pays into the burial fund and for hospitalization, the miner pays an average of $1.94 per month for medical service, and $2.56 for accident insurance. None of these fees, however, entitles him to dental care. In most instances, he also must pay an additional fee for certain vaccines, special medicines, treatments such as X-ray and radium, house calls beyond prescribed limits. Delivery of his child costs him from $10 to $50, ambulance service from $10 to $30.

"Base expenses for medical treatment, etc., average $7.74 per month. The extras bring the total average check-off per month up to an estimated $10.68 to $12.68.

"The mine workers will want to present to this conference amplification of that statement showing the conditions as they exist in detail throughout mining sections and showing that this system of medical attention is infamous from many standpoints, that it is a harsh, systematic, sordid, brutal exploitation of the mine worker, that he is the unhappy victim of this condition, and that his prayers to prolong his life and preserve the health of his family go unanswered because of his inability as an individual to rise above the grafting circumstances foisted upon him by this terrible scheme of exploitation, maintained through the years at a cost which is staggering."

"I hope that each Senator will reflect and meditate upon the terrible implications contained in this statement. Let him, if he will, make further investigation as to the truth of this indictment.

"After such consideration I cannot believe that you will vote for a bill that would restrict or prohibit workers from seeking redress of such conditions from their employer through collective bargaining contracts. It is inconceivable that such a vote could be cast that would infringe upon the fundamental right of a free citizen to contract with an employer upon the conditions under which he is willing to perform labor."[7]

7. Robert William Deimel, *The Public Relations Activity of the United Mine Workers of America Between 1946 and 1950; A Contribution to the History of Public Relations* (1971), pp. 231-33.

APPENDIX III

Comparison of Rocky Hollow Coal Company (Subsidiary of A. T. Massey) and Pittston Agreements to 1988 BCOA Agreement

Rocky Hollow Agreement	1988 BCOA Agreement	Pittston Agreement
Article I—Intent and Purpose	Article I—Enabling Clause	Article I—Enabling Clause
Same as BCOA Agreement with exception of title change and minor wording changes regarding Rocky Hollow as employer.		Same as BCOA Agreement with only minor changes in wording pertaining to Pittston as the employer.
Article IA—Scope and Coverage	Article IA—Scope and Coverage	Article IA—Scope and Coverage
Section (a) Work Jurisdiction	Section (a) Work Jurisdiction	Section (a) Work Jurisdiction
Employees will perform jobs as assigned by employer. In all cases, contracting, subcontracting, leasing and subleasing and construction work within the sole discretion of the employer.	Repair and maintenance work performed by classified employees of the employer.	Repair and maintenance work performed at the mines and facilities covered by the Agreement.

COMPARISON OF ROCKY HOLLOW AND PITTSTON AGREEMENTS TO 1988 BCOA AGREEMENT—*Continued*

Rocky Hollow Agreement	1988 BCOA Agreement	Pittston Agreement
Section (b) Exemptions Clause	Section (b) Exemptions Clause	Section (b) Exemptions Clause
Eliminates provision in BCOA Agreement relating to employer using more men than necessary for safe and efficient operation of mine.		Same as BCOA Agreement.
No person employed at other company, whether or not affiliated with Rocky Hollow, is covered under this agreement.		
Section (c) Supervisors Shall Not Perform Classified Work	Section (c) Supervisors Shall Not Perform Classified Work	Section (c) Supervisors Shall Not Perform Classified Work
Management has more discretion in working the supervisors when it is necessary than provided for in the BCOA Agreement. Under the BCOA Agreement, if a grievance arises, it shall be settled under the "settlement of disputes" section of the contract. This provision eliminated in the Rocky Hollow Agreement.		Same as BCOA Agreement.

Section (d) Management of the Mines	Section (d) Management of the Mines	Section (d) Management of the Mines
Under the Rocky Hollow Agreement, management has complete authority over the workforce.		Same as BCOA Agreement.
Section (e) Unions Rights	Section (e) Unions Rights	Section (e) Unions Rights
Rocky Hollow Agreement more restrictive than BCOA Agreement.		Same as BCOA Agreement.
Section (f) Application of This Contract to the Employers Coal Lands	Section (f) Application of This Contract to the Employers Coal Lands	Section (f) Application of This Contract to the Employers Coal Lands
Omitted from Rocky Hollow Agreement.		Insertion of Pittston Coal Group Companies (PCG Companies) instead of employers as stated in BCOA Agreement.
Section (f) Contracting and Subcontracting	Section (g) Contracting and Subcontracting	Section (g) Contracting and Subcontracting
(1) Rocky Hollow has the right to contract or subcontract without interference or restrictions	(1) Transportation of Coal—The transportation of coal may be contracted out if consistent with prior practice and custom provided that such work shall not be contracted out if any employees at the mines who perform such work are laid off.	(1) Transportation of Coal—Employer may contract out transportation of coal provided that no employee who was classified as a truck driver on or after February 1, 1988 and before April 5, 1989, is laid off or realigned.
(2) Transportation of Coal—The transportation of coal may be contracted out within sole discretion of employer.		
(3) Repair and Maintenance Work—Repair and maintenance work may not be	(2) Repair and Maintenance Work—Repair and maintenance work shall not be	(2) Repair and Maintenance Work—Employer may contract out repair and

COMPARISON OF ROCKY HOLLOW AND PITTSTON AGREEMENTS TO 1988 BCOA AGREEMENT—*Continued*

Rocky Hollow Agreement	1988 BCOA Agreement	Pittston Agreement
contracted out except where work being performed by manufacturer or supplier or where there isn't available equipment or active employees with necessary skills to perform such work. Rocky Hollow under no obligation to recall laid off employees to perform repair and maintenance work.	contracted out except where the employer does not have available equipment or regular employees (including laid off employees at the mine or central shop) to perform such work.	maintenance work of the type customarily performed by employees at the central shop.
(4) Employer may contract out rough grading work.	(3) Employer may not contract out rough grading in mine reclamation work.	(3) Same as BCOA Agreement.
(5) Where contracting out permitted, this agreement shall not limit employers choice of contractors.	(4) Where contracting out permitted, prior practice and custom shall not limit employers choice of contractors.	(4) Same as BCOA Agreement.
Section (h) Leasing, Subleasing and Licensing Out of Coal Lands	Section (h) Leasing, Subleasing and Licensing Out of Coal Lands	Section (h) Leasing, Subleasing and Licensing Out of Coal Lands
Omitted from Rocky Hollow Agreement.	(1) Changes noted in Pittston Agreement.	(1) Same as BCOA Agreement. Refers to the employer instead of employers and signatory operator as stated in BCOA Agreement.
	(2)-(7) Incorporated into Article II—Job Opportunity and Benefit Security (Jobs)	(2)-(7) Same as 1984 BCOA Agreement.

Section (g) Construction Work	Section (i) Construction Work	Section (i) Construction Work
Rocky Hollow Agreement's provision for construction work more liberal and permissive than BCOA Agreement.		Same as BCOA Agreement.
Article II—Job Opportunity and Benefit Security (Jobs)	Article II—Job Opportunity and Benefit Security (Jobs)	Article II—Job Opportunity and Benefit Security (Jobs)
Omitted from Rocky Hollow Agreement.		First paragraph deleted in Pittston Agreement and most of second paragraph.
	(A) Non-Signatory Operations—New Job Openings:	(A) Non-Signatory Operations—New Job Openings:
	1. First three out of every five job openings given to classified laid off employees.	1. The first four out of every five given to classified employees.
	2. Selection of employees for above job openings granted on basis of seniority.	2. Selection of employees taken from Preferential Recall Panel. The order of selection shall be at employers discretion.
	(3)-(6)	(3)-(6)—Same as BCOA Agreement.
	7. Same with noted exception.	7. Wording "as of date of agreement" omitted.
	8. Same with noted exception.	8. Omits wording "except as modified by Section C"

COMPARISON OF ROCKY HOLLOW AND PITTSTON AGREEMENTS TO 1988 BCOA AGREEMENT—*Continued*

Rocky Hollow Agreement	1988 BCOA Agreement	Pittston Agreement
	(B) Lessee-Licensee—Omitted from Pittston Agreement	(B) Expanded Job Opportunity Provides that contractors must offer employment for the first nineteen out of every twenty new job openings to Pittston's classified laid off employees.
Article II—Health and Safety	Article III—Health and Safety	Article III—Health and Safety
Section (a) General	Section (a) Right to a Safe Working Place	Section (a) Right to a Safe Working Place
Eliminates Safety Committee.		Same as BCOA Agreement.
Section (b) Annual Safety Equipment and Protective Clothing Allowance	Section (b) Federal Mine Safety and Health Act of 1977, as amended	Section (b) Federal Mine Safety and Health Act of 1977, as amended
Notes that if employee does not take reasonable care of equipment, or if it is lost or destroyed, he must replace it at his own cost. Protective clothing allowance $150 during each year of agreement. ($170 for the year 1988 in BCOA Agreement and $180 during 1989 through 1992)	Omitted from Rocky Hollow Agreement.	Same as BCOA Agreement.

Section (c) Physical Examination	Section (c) Joint Industry Health and Safety Committee	Section (c) Joint Health and Safety Committee
(1) Same as BCOA Agreement. (2) Within sole discretion of employer to decide when physical examination required. (3) Corresponds to number (2) in BCOA Agreement with the exception that if employee has impairment, he shall not be allowed to return to work if it *prevents him from performing his regular work.* (4) Employee not entitled to any pay when eligible to return from a strike for a physical examination. (5) Same as number (3) in BCOA Agreement. (6) Same as number (4) in BCOA Agreement.		Same as BCOA Agreement except the word "industry" has been deleted from title and mention in section.
Section (d) Safety Rules and Regulations	Section (d) Mine Health and Safety Committee	Section (d) Mine Health and Safety Committee
Adds that rules and regulations of the employer complied with *at all times.* Omits second paragraph of BCOA Agreement pertaining to implementation of revised safety rule or regulation.	Omitted from Rocky Hollow Agreement.	Same as BCOA Agreement.

COMPARISON OF ROCKY HOLLOW AND PITTSTON AGREEMENTS TO 1988 BCOA AGREEMENT—*Continued*

Rocky Hollow Agreement	1988 BCOA Agreement	Pittston Agreement
Section (e) Access to Mine	Section (e) Access to Mine	Section (e) Access to Mine
Restricted to statutory rights under federal or state laws.	(1) Changes noted in Pittston Agreement.	(1) Requires advance notice of visit to mine by representative of the UMWA Department of Occupational Health and Safety. In addition, if the employer does not participate, union representative will not be denied access.
	(2)-(5)	(2)-(5) Same as BCOA Agreement.
Section (f) Reports	Section (f) Reports	Section (f) Reports
Elimination of Mine Health and Safety Committee. Records limited to requirements of Federal Mine Safety and Health and Administration.		Same as BCOA Agreement.
Section (g) Minimum Age	Section (g) Safety Rules and Regulations	Section (g) Safety Rules and Regulations
Lists Rocky Hollow mine instead of any mine, and omits section in BCOA Agreement with reference to state law providing a higher minimum age.	Corresponds to Section (d) in Rocky Hollow Agreement.	Same as BCOA Agreement.

Section (h) Workers Compensation and Occupational Disease	Section (h) Cooperation in Development of Mining Plans	Section (h) Cooperation in Development of Mining Plans
Rocky Hollow employees provided protection under applicable laws. Eliminates section in BCOA Agreement regarding refusal to carry out this directive.	Omitted from Rocky Hollow Agreement.	Same as BCOA Agreement.
	Section (i) Preservation of Individual Safety Rights	Section (i) Preservation of Individual Safety Rights
	Omitted from Rocky Hollow Agreement.	Same as BCOA Agreement.
	Section (j) Physical Examination	Section (j) Physical Examination
	Corresponds to Section (c) in Rocky Hollow Agreement.	Same as BCOA Agreement.
	Section (k) Minimum Age	Section (k) Minimum Age
		Same as BCOA Agreement.
	Section (l) Workers Compensation and Occupational Disease	Section (l) Workers Compensation and Occupational Disease
	Same with noted exceptions. Corresponds to Section (h) in Rocky Hollow Agreement.	Same as BCOA Agreement with minor wording changes to apply to Pittston Coal Group as the employer

COMPARISON OF ROCKY HOLLOW AND PITTSTON AGREEMENTS TO 1988 BCOA AGREEMENT—*Continued*

Rocky Hollow Agreement	1988 BCOA Agreement	Pittston Agreement
	Section (m) Safety Equipment and Protective Clothing Allowance	Section (m) Safety Equipment and Protective Clothing Allowance
	$180 for 1989, 1990, 1991 and 1992. Corresponds to Section (b) in Rocky Hollow Agreement.	Same as BCOA Agreement—$180 for each year of Agreement.
	Section (n) Maintenance	Section (n) Maintenance
	Omitted in Rocky Hollow Agreement.	Same as BCOA Agreement.
	Section (o) Special Safety Problem Areas	Section (o) Special Safety Problem Areas
	Omitted in Rocky Hollow Agreement.	Same as BCOA Agreement with numbers 13 and 14 added to Pittston Agreement pertaining to wearing of eye protection and hearing protection.
	Section (p) Settlement of Health or Safety Disputes	Section (p) Settlement of Health or Safety Disputes
	Omitted in Rocky Hollow Agreement.	Same as BCOA Agreement.
Article III—Wages and Hours	Article IV—Wages and Hours	Article IV—Wages and Hours
Section (a) Basic Work Week	Section (a) Basic Work Week	Section (a) Basic Work Week
Employer has the right to set basic work week.		Same as BCOA Agreement.

Section (b) Basic Work Day	Section (b) Basic Work Day	Section (b) Basic Work Day
1. Basic Work Day—nine hours for all employees regardless of portal-to-portal. Employees paid only if they actually work nine hours portal-to-portal. Face employees shall work eight hours. Overtime paid only for hours worked in excess of forty hours per basic work week, with no pyramiding of overtime. 2. Employer has right to change basic work day to more or less than nine hours.	(1)-(3) Numbers 1-3 all contained in number 1 in Rocky Hollow Agreement.	(1)-(3) Same as BCOA Agreement.
Section (c) Lampman	**Section (c) Lampman**	**Section (c) Lampman**
Omitted from Rocky Hollow Agreement.		Same as BCOA Agreement.
Section (c) Saturday, Sunday and Premium Work	**Section (d) Saturday, Sunday and Premium Work**	**Section (d) Saturday, Sunday and Premium Work**
1. Work performed on Saturday and Sunday shall be paid at regular rate. Work on holiday paid for at time and one-half except work performed on Thanksgiving, Christmas or Easter which shall be paid at double time or double rates. No triple time paid under any circumstances.	(1)-(7)	(1)-(7) Same as BCOA Agreement.

COMPARISON OF ROCKY HOLLOW AND PITTSTON AGREEMENTS TO 1988 BCOA AGREEMENT—*Continued*

Rocky Hollow Agreement	1988 BCOA Agreement	Pittston Agreement
2. Idle day work equally shared where applicable. Corresponds to number seven in BCOA Agreement in which idle day work must be equally shared in accordance with past practice and custom.		
3. All employees with necessary qualifications given fair and equal opportunity to work. BCOA reference to laying off employees for purpose of denying them six days work omitted.		
4. Employees given equal opportunity to share overtime work. BCOA: employees shall be given opportunity to work same number of days per week as the number of days mine produces coal and given equal opportunity to share available work on premium days.		
5. Same as BCOA Agreement—Notification of idle day work.		
6. Employee may refuse idle day work if less than regular rate—BCOA Agreement: regular *classified* rate.		

Section (d) Standard Hourly Wage Rate Provides four classifications: $15.15—Electricians $14.60—Underground Face $13.80—Underground Outby $13.60—Outside Workers	Section (e) Standard Daily Wage Rate Five pay grades in deep mines, four pay grades in preparation plants and other surface facilities and five pay grades in strip and auger mines. Employees paid in accordance with standard daily wage rates as set forth in Appendix A and the job titles within classifications grouped in Appendix A, Part I.	Section (e) Standard Daily Wage Rate Same as BCOA Agreement.
Section (e) Wage Increases 10/17/89—$0.20 per hour 10/17/90—$0.15 per hour BCOA Agreement: Effective Date — Wage Increase 2/1/88 — $0.25 per hour 2/1/89 — $0.35 per hour 2/1/90 — $0.45 per hour		
Article IV—Helpers on Face Equipment in Underground Mines Helpers eliminated unless required by law.	Article V—Helpers on Face Equipment in Underground Mines	Article V—Helpers on Face Equipment in Underground Mines Same as BCOA Agreement.

COMPARISON OF ROCKY HOLLOW AND PITTSTON AGREEMENTS TO 1988 BCOA AGREEMENT—*Continued*

Rocky Hollow Agreement	1988 BCOA Agreement	Pittston Agreement
Article V—Shifts and Shift Differentials	Article VI—Shifts and Shift Differentials	Article VI—Shifts and Shift Differentials
Section (a) Multiple Shifts	Section (a) Multiple Shifts	Section (a) Multiple Shifts
Employer has right to work seven days a week, twenty-four hours a day with crews as employer may elect in its sole discretion from time to time.		Same as BCOA Agreement.
Section (b) Hoisting of Coal	Section (b) Hoisting of Coal	Section (b) Hoisting of Coal
Same as BCOA Agreement.		Same as BCOA Agreement.
Section (c) Shift Differentials	Section (c) Shift Differentials	Section (c) Shift Differentials
Afternoon: $.25 for each additional hour worked. Midnight: $.35 for each hour worked.	Afternoon: $.30 by the day or by the ton. Midnight: $.40 by the day or by the ton.	Same as BCOA Agreement.
Section (d) Working into Next Shift	Section (d) Working into the Next Shift	Section (d) Working into the Next Shift
Same as BCOA Agreement.		Same as BCOA Agreement.
Section (e) Call Back	Section (e) Call Back	Section (e) Call Back
Paid time and one-half only after forty hours worked in basic work week.		Same as BCOA Agreement.
Section (f) Shift Rotation	Section (f) Shift Rotation	Section (f) Shift Rotation
Provides exception after forty hours of work in basic work week.		Same as BCOA Agreement.

Article VI—Mine Communication Committees	Article VII—Mine Communication Committees	Article VII—Mine Communication Committees
Safety Committee chairman omitted from Mine Communication Committee. Also eliminates tour of mine by International Union together with employer.		Same as BCOA Agreement with the exception of insertion of word "designated." Employer-*designated* representative shall tour the mine.
Article VII—Shift Start/End Time	Article VIII—Starting Time	Article VIII—Starting Time
Starting time and duration of each work shift exclusively established and changed by employer.	Section (a) Shift Starting Time	Section (a) Shift Starting Time
	Omitted in Rocky Hollow Agreement.	Same as BCOA Agreement.
	Section (b) Lowering Employees	Section (b) Lowering Employees
	Omitted in Rocky Hollow Agreement.	Same as BCOA Agreement.
	Section (c) Safety and Maintenance	Section (c) Safety and Maintenance
	Omitted in Rocky Hollow Agreement.	Same as BCOA Agreement.
	Section (d) Surface Facilities	Section (d) Surface Facilities
	Omitted in Rocky Hollow Agreement.	Same as BCOA Agreement.
	Section (e) Crew Changes	Section (e) Crew Changes
	Omitted in Rocky Hollow Agreement.	Same as BCOA Agreement.
Article VIII—Allowances	Article IX—Allowances	Article IX—Allowances
Section (a) Bereavement	Section (a) Bereavement Pay	Section (a) Bereavement Pay
Same as BCOA Agreement.		Same as BCOA Agreement.

COMPARISON OF ROCKY HOLLOW AND PITTSTON AGREEMENTS TO 1988 BCOA AGREEMENT—*Continued*

Rocky Hollow Agreement	1988 BCOA Agreement	Pittston Agreement
Section (b) Jury Duty	Section (b) Jury Duty	Section (b) Jury Duty
Same as BCOA Agreement.		Same as BCOA Agreement.
Section (c) Reporting Pay	Section (c) Reporting Pay	Section (c) Reporting Pay
Same as BCOA Agreement.		Same as BCOA Agreement.
Section (d) Military Duty	Section (d) Military Duty	Section (d) Military Duty
Same as BCOA Agreement.		Same as BCOA Agreement.
Section (e) Personal or Sick Leave	Section (e) Personal or Sick Leave	Section (e) Personal or Sick Leave
1. Any employee who has worked a year or more with Rocky Hollow shall be eligible for *two* working days for personal or sick leave.	1. BCOA Agreement provides *five* working days for personal or sick leave.	1. Same as BCOA Agreement with exception of last year of Pittston Agreement, 1994, in which each classified employee shall be eligible for *three* personal or sick leave days.
2. Same as BCOA Agreement.	2.	2. Same as BCOA Agreement.
3. Employee must notify supervisor at least seven calendar days in advance when he plans to take personal or sick leave.	3. Employee notifies supervisor twenty-four hours in advance when he is unable to report.	3. Same as BCOA Agreement.
4. Same as BCOA Agreement.	4.	4. Same as BCOA Agreement.
5. Same as BCOA Agreement.	5.	5. Same as BCOA Agreement.

5. Subsection (a)—Omitted from Rocky Hollow Agreement.	5. Subsection (a)—*Employer Options on Personal or Sick Leave Pay* Employer may charge up to two personal and sick leave days if an employee is absent from work and has not requested a personal or sick leave day.	5. Subsection (a)—Omitted. Listed as number 6. The employer may charge up to five personal or sick leave days.
5. Subsection (b)—Omitted from Rocky Hollow Agreement 6. Employer pays personal or sick leave pay to employee who has not exhausted his pro-rata share of such days. 7. New to Rocky Hollow Agreement. Provides in 1991, prior to expiration of agreement, employee shall be entitled to one personal or sick leave day. Any other personal or sick leave days will be determined by successor agreement.	5. Subsection (b)—*Employee Options*	5. Subsection (b)—Omitted in Pittston Agreement.
Article IX—Wage Increase	Article X—Wage Increase	Article X—Annual Wage Increase
Part of Article III—Wages and Hours, Section (e) of Rocky Hollow Hollow Agreement.	Effective Date — Wage Increase 2/1/88 — $0.25 per hour 2/1/89 — $0.35 per hour 2/1/90 — $0.45 per hour	Effective date of agreement, $0.40 per hour and on the anniversary date of agreement in 1991 and 1992, $0.40 per hour, total of $1.20 per hour. In addition, if amount received as health care reimbursement benefit exceeds actual covered medical expenses incurred

COMPARISON OF ROCKY HOLLOW AND PITTSTON AGREEMENTS TO 1988 BCOA AGREEMENT—*Continued*

Rocky Hollow Agreement	1988 BCOA Agreement	Pittston Agreement
		by eligible participants during applicable deductible period, then participant retains the excess as a supplemental compensation benefit.
Article IX—Sickness and Accident Benefits	Article XI—Sickness and Accident Benefits	Article IX—Sickness and Accident Benefits
Section (a) General Purpose	Section (a) General Purpose	Section (a) General Purpose
Provides that no other entity other than Rocky Hollow will be responsible for any obligation.		Same as BCOA Agreement.
Section (b) Eligibility	Section (b) Eligibility	Section (b) Eligibility
Eliminates "light duty" provision contained in BCOA Agreement. Also provides that "classified employment" means employment in a classified job which hasn't been broken by violation of Article XXII—no strike/no lockout clause—which is a new article not contained in the BCOA Agreement.		Same as BCOA Agreement.

Section (c) Commencement and Duration of Benefits	Section (c) Commencement and Duration of Benefits	Section (c) Commencement and Duration of Benefits
Same as BCOA Agreement.		Same as BCOA Agreement.
Section (d) Amount and Payment of Benefits	Section (d) Amount and Payment of Benefits	Section (d) Amount and Payment of Benefits
Provides $165 during first year of agreement; $175 during second year; and $185 during each subsequent year of agreement.	Provides $200 per week during first year of agreement; $210 per week for second year of agreement and $220 per week for third year, fourth and fifth years of agreement.	Sickness and accident benefits payable: $210 during first year of agreement. Subsequent years of agreement: $220 per week.
Section (e) Filing of Claims for Benefits	Section (e) Filing of Claims for Benefits	Section (e) Filing of Claims for Benefits
Same as BCOA Agreement.		Same as BCOA Agreement.
Section (f) Structure and Administration	Section (f) Structure and Administration	Section (f) Structure and Administration
The operation of the sickness and accident benefits plan sole responsibility of Rocky Hollow. BCOA Agreement: joint responsibility between employers and union. Rocky Hollow or its insurance carrier shall investigate employees claims. Under BCOA Agreement, trustees of the UMWA Health and Retirement Fund prepares list of acceptable insurance carriers. Rocky Hollow chooses own insurance carrier.	Joint responsibility of employers and union.	Sole responsibility of Pittston. Pittston may also designate insurance carrrier. The union shall have the right to investigate disputed claims. Also, the union shall advise the employer of the results of the investigation and may request employer to take steps to remedy any inconsistencies. If any dispute remains unresolved, it shall be subject to arbitration.

COMPARISON OF ROCKY HOLLOW AND PITTSTON AGREEMENTS TO 1988 BCOA AGREEMENT—*Continued*

Rocky Hollow	BCOA	Pittston
Article X—Holidays	Article XII—Holidays	Article XII—Holidays
Section (a) Holidays Observed	Section (a) Holidays Observed	Section (a) Holidays Observed
Provides for ten paid holidays. Eliminates eight-hour holiday and the employee's birthday. No holidays observed following expiration of agreement, unless granted in successor agreement.	Provides for eleven paid holidays including the employee's birthday.	Same as BCOA Agreement with exception of last year of Pittston contract. In 1994, the fifth year of the agreement, only three holidays will be observed: eight-hour holiday, Memorial Day and employee's birthday.
Section (b) Sunday Holidays	Section (b) Sunday Holidays	Section (b) Sunday Holidays
Eliminates birthday holiday.		Same as BCOA Agreement.
Section (c) Monday Holidays	Section (c) Monday Holidays	Section (c) Monday Holidays
Omitted in Rocky Hollow Agreement.		Same as BCOA Agreement.
Section (c) Pay for Holidays Worked	Section (d) Pay for Holidays Worked	Section (d) Pay for Holidays Worked
Paid only time and one-half for all hours worked on holidays with the exception of Thanksgiving Day, Christmas and Easter which will be paid double time or double time rates.		Same as BCOA Agreement.
Section (d) Pay for Holidays Not Worked	Section (e) Pay for Holidays Not Worked	Section (e) Pay for Holidays Not Worked
Eliminates regularly scheduled overtime rates provided in BCOA Agree-		Same as BCOA Agreement.

ment. Also provides that employee did not incur unexcused absence on his last scheduled day. Any employee who is on strike or a work stoppage shall not receive holiday pay. (New to Rocky Hollow.)		
Section (e) Holidays During Vacation Period	Section (f) Holidays During Vacation Period	Section (f) Holidays During Vacation Period
Eliminates birthday holiday. No addi tional pay for designated holiday employee wishes to take.		Same as BCOA Agreement.
Section (g) Birthday Holidays	Section (g) Birthday Holidays	Section (g) Birthday Holidays
Omitted in Rocky Hollow Agreement.		Same as BCOA Agreement.
Section (f) Holidays for Sick and Injured	Section (h) Holidays for Sick and Injured	Section (h) Holidays for Sick and Injured
Same as BCOA Agreement.		Same as BCOA Agreement.
Section (g) Time of Payment	Section (i) Time of Payment	Section (i) Time of Payment
Same as BCOA Agreement.		Same as BCOA Agreement.
Article XI—Vacation	Article XIII—Regular Vacation	Article XIII—Regular Vacation
Section (a) Annual Vacation	Section (a) Annual Vacation	Section (a) Annual Vacation
To be determined by employer and employer may establish separate vacation periods and assign employees to those periods.		Same as BCOA Agreement.

COMPARISON OF ROCKY HOLLOW AND PITTSTON AGREEMENTS TO 1988 BCOA AGREEMENT—*Continued*

Rocky Hollow Agreement	1988 BCOA Agreement	Pittston Agreement
Section (b) Dates of Regular Vacation Period	Section (b) Dates of Regular Vacation Period	Section (b) Dates of Regular Vacation Period
Employer designates fourteen-day period in June, July or August.	Lists time and dates.	Lists time and last Saturday in June to second Saturday in July, etc.
Section (c) Staggered Regular Vacation	Section (c) Staggered Regular Vacation	Section (c) Staggered Regular Vacation
Corresponds to Section (e) of Rocky Hollow Agreement.	No more than fifteen percent of work force at mine elects to be off on the same day.	No more than *ten percent* elects to be off on the same day.
Section (c) Qualifying Period and Amount of Vacation Payment Under This Agreement	Section (d) Qualifying Period and Amount of Payment	Section (d) Qualifying Period and Amount of Payment
1. Paid *ten* times the employee's day wage rate. Doesn't specify years as BCOA Agreement does. For example, June 1, 1989, to May 31, 1990, etc.	1. Qualifying Period same with exception of dates. Begins with June 1, 1987, to May 31, 1988.	1. Only dates are different. Pittston begins with June 1, 1989, to May 31, 1990.
2. *Shift Differentials: All Three Shifts* \$0.25—Afternoon Shift \$0.35—Midnight Shift Divide total of \$0.60 by three for an average of \$0.20 per hour.	2. *Shift Differentials* \$0.30—Afternoon Shift \$0.40—Midnight Shift Divide total of \$0.70 by three for an average of 23 1/3¢ per hour.	2. *Shift Differentials* Same as BCOA Agreement.

For Afternoon and Day Shifts $0.25 divided by two to arrive at average of 12.5¢ an hour.	*For Afternoon and Day Shifts* $0.30 divided by two to arrive at average of 15¢ an hour.	*For Afternoon and Day Shifts* Same as BCOA Agreement.
3. *Sick and Injured Employees* All terms and provisions carried forward for one year. No payment if employee engages in strike or receives strike benefits at termination of agreement.	3. *Sick and Injured Employees*	3. *Sick and Injured Employees* Same as BCOA Agreement.
4. *Pro-Rata Payments* Same as BCOA Agreement.	4. *Pro-Rata Payments*	4. *Pro-Rata Payments* Same as BCOA Agreement.
Section (e) Floating Vacation Days	Section (e) Floating Vacation Days	Section (3) Floating Vacation Days
Corresponds to Section (j) in Rocky Hollow Agreement.	Provides for no floating vacation days from January 1, 1993 until expiration of agreement.	Provides two floating vacation days for 1994 until expiration of agreement.
	Declaration for floating vacation days pertaining to 1988 and 1993.	1990 and 1994.
	1. Daily Basis	1. Daily Basis—Same as BCOA Agreement.
	2. Christmas Shutdown Period: Same for 1990, 1992 and 1992	2. Christmas Shutdown Period: During 1993, four floating vacation days used with Christmas Eve holiday.
Section (d) Time of Payment	Section (f) Time of Payment	Section (f) Time of Payment
Same as BCOA Agreement.		Same as BCOA Agreement.

COMPARISON OF ROCKY HOLLOW AND PITTSTON AGREEMENTS TO 1988 BCOA AGREEMENT—*Continued*

Rocky Hollow Agreement	1988 BCOA Agreement	Pittston Agreement
Section (e) Staggered Summer Vacation Period		
Does away with past practice. UMWA notification eliminated if employer elects to operate mine without interruption. Provides that no more than ten percent of work force not off on same day. (BCOA Agreement provides for fifteen percent.) If conflict arises over choice of vacation, it will be determined *first* based upon needs of employer, *second* on a seniority basis.		
Institution of staggered vacation period under control of employer and employees will be notified by May 15 of the year in which vacation will fall.		
Section (f) Christmas Vacation Period		
Corresponds to Section (e) in BCOA Agreement. 1. Provides additional three days to coincide with Christmas. Such days		

scheduled by employer. BCOA Agreement provides for four specific additional days. Dates omitted from Rocky Hollow Agreement. Referred to as floating vacation days. 2. Employees compensated at regular day-wage rate. 3. Employees notified by September 30 when vacation to occur. In BCOA Agreement, notice must be given by November 15.		
Section (g) Obligation of Payment	Section (g) Obligation of Payment	Section (g) Obligation of Payment
Eliminates section in BCOA Agreement in reference to selling of mine, leasing or subleasing.		Same as BCOA Agreement.
Section (h) Work During Vacation	Section (h) Work During Shutdown	Section (h) Work During Shutdown
Eliminates provision for work at coke plants. Management has the right to operate during vacation periods.		Same as BCOA Agreement.
Section (i) Vacation Pursuant to Successor Agreement		
Any vacation period observed after expiration of agreement shall be determined solely by provision of successor		

Rocky Hollow Agreement	1988 BCOA Agreement	Pittston Agreement
agreement. (New to Rocky Hollow Agreement.)		
Section (j) Pay in Lieu of Floating Vacation Days Under This Agreement		
Corresponds to Section (e), number 2, in BCOA Agreement.		
Paid at regular rate for pro-rata share of floating vacation days. Eliminates inclusion of regularly scheduled overtime. Must work four years longer at Rocky Hollow to pick up additional vacation days. From six years but less than seven years for one day in BCOA Agreement to ten years and less than eleven years for one day in Rocky Hollow Agreement, etc.		
Article XII—Graduated Vacation	Article XIV—Graduated Vacation	Article XIV—Graduated Vacation
Section (a) General	Section (a) General	Section (a) General
Omits six years to less than ten years in BCOA Agreement.		Same as BCOA Agreement.

Additional Vacation Days: 10 years but less than 11—1 11 years but less than 12—2 12 years but less than 13—3 13 years but less than 14—4 14 years but less than 15—5 15 years or over . . . . —6 Omits 16 years but less than 18 years and 18 years or over.	Additional Vacation Days: 10 years but less than 11— 5 11 years but less than 12— 6 12 years but less than 13— 7 13 years but less than 14— 8 14 years but less than 15— 9 15 years but less than 16—10	
Section (b) Definition of Additional Days	Section (b) Definition of Additional Days	Section (b) Definition of Additional Days
Same as BCOA Agreement.		Same as BCOA Agreement.
Section (c) Definition of Continuous Employment	Section (c) Definition of Continuous Employment	Section (c) Definition of Continuous Employment
1. Adds layoff and violation of no-strike clause to definition of "continuous employment." 2. Employee's employment at active mine operations subject to reciprocal agreement. (New to Rocky Hollow Agreement.)	1. Continuous employment not interrupted by layoff, transfer and sale and subleasing of mine. (2)-(4) Omitted from Rocky Hollow Agreement.	(1)-(4) Same as BCOA Agreement.
Section (d) Amount of Continuous Employment	Section (d) Amount of Continuous Employment	Section (d) Amount of Continous Employment
1. Based upon settlement made with UMW employees. Numbered (1) and (2)	1. Same with noted exceptions.	1. Same as BCOA Agreement with exception of dates. BCOA 1988, Pitts-

COMPARISON OF ROCKY HOLLOW AND PITTSTON AGREEMENTS TO 1988 BCOA AGREEMENT—*Continued*

Rocky Hollow Agreement	1988 BCOA Agreement	Pittston Agreement
in BCOA Agreement. Number (2) relating to graduated vacation in calendar year 1988 omitted. Includes provision for period from October 1, 1984, through the employee's return to work. This period will not be counted as continuous employment, except for discharge and returning to work as a result of settlement.	2. Omitted from Pittston Agreement.	ton 1990. Also adds that employees recalled to active employment shall only be entitled to graduated vacation days on a pro-rata basis equal to number of months such employees on payroll during the calendar year of their recall.
Section (e) Time of Payment	Section (e) Time of Payment	Section (e) Time of Payment
Same as BCOA Agreement.		Same as BCOA Agreement.
Section (f) Rate of Payment	Section (f) Rate of Payment	Section (f) Rate of Payment
Same as BCOA Agreement.		Same as BCOA Agreement.
Section (g) Scheduling and Pay in Lieu	Section (g) Scheduling and Pay in Lieu	Section (g) Scheduling and Pay in Lieu
Employer must have three months notice from employee on graduated vacation day.	Same with noted exception.	Provides that not more than ten percent of work force scheduled off at any time.
Section (h) Sick and Injured Employees	Section (h) Sick and Injured Employees	Section (h) Sick and Injured Employees
Same as BCOA Agreement.		Same as BCOA Agreement.

Article XIII—Checkoff	Article XV—Checkoff	Article XV—Checkoff
Employer must be given written authorization by employee to checkoff any payments, and employee is under no obligation to have checkoff.		Same as BCOA Agreement with exception of adding sentence stating "upon execution of this agreement and only during the term of this agreement" at the beginning of the article.
Article XIV—Training	Article XVI—Training	Article XVI—Training
Section (a) Priority	Section (a) Priority	Section (a) Priority
Omitted from Rocky Hollow Agreement.	Section pertaining to employers operating (three) or fewer mines omitted from Pittston Agreement. Trained staff representative to devote "full time" in assisting Joint Industry Training Committee.	Omits word "industry" from Joint Industry Training Committee. Three representatives of management instead of industry. The committee shall consist of the International Union and "management" instead of BCOA assigning trained staff representative and only *as needed* to assist *Joint Training Committee.*
Section (a) Orientation for New Employees	Section (b) Orientation for New Employees	Section (b) Orientation for New Employees
1. Same as BCOA Agreement. 2. Same as BCOA Agreement. 3. Subsection (i), same as BCOA Agreement.	(1)-(3), subsections (i)-(vi) same in Pittston Agreement. (viii)-(xiv)—same in Pittston Agreement.	(3) Subsection (vii)—Adds eye and hearing protective devices required by agreement.

COMPARISON OF ROCKY HOLLOW AND PITTSTON AGREEMENTS TO 1988 BCOA AGREEMENT—*Continued*

Rocky Hollow Agreement	1988 BCOA Agreement	Pittston Agreement
3. Subsection (ii)—Authority and safety responsibilities of the union, not the Mine Health and Safety Committee as contained in the BCOA Agreement. 3. Subsections (iii)-(xiv), same as BCOA Agreement. 4. Subsection (i)—same as BCOA Agreement. 4. Subsection (ii)—Replaces Mine Health and Safety Committee with "Union." 4. Subsection (iii)-(x)—same as BCOA Agreement. 5. Notes program at Rocky Hollow. 6. Same as BCOA Agreement.	(4) Subsections (iv), (x)—same in Pittston Agreement. (4) Subsections (i), (ii)—same in Pittston Agreement. (5) & (6)—same in Pittston Agreement.	(4) Subsection (iii)—Adds eye and hearing protective devices as required by agreement.
Section (b) General Retraining Programs	Section (c) General Retraining Programs	Section (c) General Retraining Programs
1. Each program subject to modification by personnel designated by employer. 2. Same as BCOA Agreement. 3. Subsections (i) and (ii)—same as BCOA Agreement. 3. Subsections (iii)—Instruction regarding procedures for escape and *ventila-*	1. Lists years 1988 through 1992 for participation in general retraining program. 2. Subsections (i)-(x) same in Pittston Agreement. 3. Subsections (i)-(xv) same in Pittston Agreement.	1. Omits years, states "each calendar year." 2. Subsection (xi) (New to Pittston Agreement)—Instruction in use of eye and hearing protective devices. 3. Subsection (xvi) (New to Pittston Agreement)—Instruction in use of eye and hearing protective devices.

tion. In BCOA Agreement, escape and evacuation. 3. Subsections (iv)-(xv)—same as BCOA Agreement. 4. Same as BCOA Agreement.		4. Same as BCOA Agreement.
Section (c) Safety Training for Specific Job	Section (d) Safety Training for Specific Job	Section (d) Safety Training for Specific Job
Same as BCOA Agreement.		Same as BCOA Agreement.
Section (d) Maintenance Training	Section (e) Maintenance Training and Rate of Pay	Section (e) Maintenance Training and Rate of Pay
Employer designates representatives for training, and pay grades are different since Rocky Hollow classifications fall within four categories.	Same with noted exception.	Same as BCOA Agreement and adds Part V and VI to maintenance job covered by Appendix A, Parts II and III.
Section (e) New Inexperienced Employees	Section (f) New Inexperienced Employees at Underground Mines	Section (f) New Inexperienced Employees at Underground Mines
Same as BCOA Agreement.		Same as BCOA Agreement.
Section (f) General Training Provisions	Section (g) General Training Provisions	Section (g) General Training Provisions
1. Same as BCOA Agreement. 2. Designated union representative employed at Rocky Hollow mine shall review each training program. BCOA Agreement: local and district officials of the union (including Mine Health &	(1)-(4)	(1)-(4) Same as BCOA Agreement.

Rocky Hollow Agreement	1988 BCOA Agreement	Pittston Agreement
Safety Committee) shall review training program. 3. Determination of trainability within sole discretion of employer. Omits definition of trainability. 4. If employee fails to complete training program, he shall be given a job he is qualified to perform. If none exists, he will be placed on the mine panel. BCOA Agreement: if employee fails, he is allowed to return to his prior job. If it is not available, then he is assigned to the general labor crew.		
Article XV—Seniority	Article XVII—Seniority	Article XVII—Seniority
Section (a) Definition of Seniority	Section (a) Definition of Seniority	Section (a) Definition of Seniority
1. It is up to employer to determine who is the most qualified individual for a job.		Same as BCOA Agreement.
2. Panel rights shall apply at other mines wholly owned and operated by Rocky Hollow. 3. Grants opportunity to revise panel form.	Applies to Reciprocal Agreement contained in Rocky Hollow Agreement.	

Section (b) Reduction; Realignment	Section (b) Reduction; Realignment	Section (b) Reduction; Realignment
1. *Reduction in Work Force:* Employees retained in accordance of sole discretion of employer.	1. *Reduction in Work Force:* The reduction of the work force done on a seniority basis.	1. *Reduction in Work Force:* Same as BCOA Agreement.
2. *Realignment Procedure:* (a) Subsections (i) and (ii)—Same as BCOA Agreement.	2. *Realignment Procedure* (a) Subsections (i) and (ii)	2. *Realignment Procedure:* (a) Subsections (i) and (ii)—Same as BCOA Agreement.
(b) Employees assigned available jobs if displaced from their job title in accordance of sole discretion of employer. (b) (i)—Same as BCOA Agreement.	(b) (i)	(b) (i)—Same as BCOA Agreement.
(b) (ii)—Stipulates that employees who step in and perform available work must be *most qualified.*	(b) (ii)—Employees who have ability perform available work.	(b) (ii)—Same as BCOA Agreement.
(b) (iii)—*Shift Assignment:* Eliminates reference to subparagraph (1) which has reference to senior employees.	(b) (iii)	(b) (iii)—Same as BCOA Agreement.
(c) Provides for sole discretion of employer in determining seniority.	(c)	(c) Same as BCOA Agreement.
(d) Provides for sole discretion of employer in determining seniority. Also, employee must be determined by a licensed physician to be able to perform such job and return to work within five days after realignment.	(d) Omits requirement for licensed physician.	(d) Same as BCOA Agreement.

COMPARISON OF ROCKY HOLLOW AND PITTSTON AGREEMENTS TO 1988 BCOA AGREEMENT—*Continued*

Rocky Hollow Agreement	1988 BCOA Agreement	Pittston Agreement
(e) Employee request for other shift preference followed only when practicable. (f) Management retains right to maintain facemen and electricians in regular jobs. (New to Rocky Hollow Agreement.)	(e) Employee request for shift preference "shall be followed."	(e) Same as BCOA Agreement.
Section (c) Layoff Procedure	Section (c) Layoff Procedure	Section (c) Layoff Procedure
Mine Committee shall lodge exceptions to management. Management shall review exceptions and determine whether to modify its plan. (New to Rocky Hollow Agreement.)		Same as BCOA Agreement with exception of listing of "the PCG Companies" mines.
On panel form, employee must list previous *verifiable* mining experience with other employers. Also has reference to reciprocal agreement which is part of Rocky Hollow Agreement and being placed on its panel. Failure by employee to file panel form in five-day period shall result in forfeiture of rights of recall.	Only requests previous mining experience.	

Section (d) Panels	Section (d) Panels	Section (d) Panels
Duration of agreement only will employees be placed on a panel. A panel member shall not be considered for any jobs not listed on panel form. Name will be taken off panel if not recalled to work within five years. (Not contained in BCOA Agreement.)	Does not make exception for jobs not listed on panel form.	Same as BCOA Agreement.
Section (e) Panel Custodians	Section (e) Panel Custodians	Section (e) Panel Custodians
Mine committee chairman and superintendent joint custodians of panel records.	Superintendent and recording secretary joint custodians of panel records.	Same as BCOA Agreement.
Five days to report to work after acceptance of notice of recall.	Reasonable time to report to work.	
Reciprocal Agreement section added. (New to Rocky Hollow Agreement.)		
Section (f) Panel Members Accrue Seniority	Section (f) Panel Members Accrue Seniority	Section (f) Panel Members Accrue Seniority
Omits section in BCOA Agreement providing for laid-off employees not accruing any seniority if they place their names on a panel at other mines.		Same as BCOA Agreement.
Section (g) Right to be Recalled	Section (g) Right to be Recalled	Section (g) Right to be Recalled
Return to work in accordance with terms		Same as BCOA Agreement.

COMPARISON OF ROCKY HOLLOW AND PITTSTON AGREEMENTS TO 1988 BCOA AGREEMENT—*Continued*

Rocky Hollow Agreement	1988 BCOA Agreement	Pittston Agreement
of section (e) and if employee does not return to work, he shall sacrifice his seniority and recall rights at employer's operation, and his name will be removed from the panel list.		
Section (h) Recall of Persons on Layoff Status	Section (h) Recall of Persons on Layoff Status	Section (h) Recall of Persons on Layoff Status
Omitted from Rocky Hollow Agreement.	First paragraph same with noted exception. 1. 2. and 3. Same with noted exception.	Job or training vacancy at a mine "of the PCG Companies." 1. Same as BCOA Agreement. 2. and 3. Lists PCG Companies.
Section (h) Job Bidding	Section (i) Job Bidding	Section (i) Job Bidding
1. Seven days no less than five for posting of job or vacancy. 2. Under sole discretion of employer to determine who is most qualified for existing job. 3. Five days after posting period employee selected for job or vacancy. Successful bidder may be asked to remain in present job for a reasonable period of time.	1. Five days not less than three for posting of job or vacancy. 2. 3. Three days after end of posting period employee selected for job or vacancy. Successful bidder may remain in present job for a period not to exceed thirty days.	(1)-(10) Same as BCOA Agreement.

4. Same as BCOA Agreement.		
5. Same as BCOA Agreement.		
6. Omits reference to senior employee making a bid for a job.		
7. Same as BCOA Agreement.		
8. Omitted in Rocky Hollow Agreement exception for no eligible bidders and job being awarded to senior bidder. Also omits provisions for bidding from an inside job to an outside job and to bid from a job to which employee was realigned.		
9. Omitted from Rocky Hollow Agreement provision that job has been properly identified.		
10. Same as BCOA Agreement.		
11. Permanent vacancies in machine operator positions awarded in accordance with section (a) definition of seniority.		
12. New to Rocky Hollow Agreement. Has reference to Reciprocal Agreement to Rocky Hollow contract.	Not contained in BCOA Agreement.	
Section (i) Training for Vacancy Not Filled by Bidding	Section (j) Training for Vacancy Not Filled by Bidding	Section (j) Training for Vacancy Not Filled by Bidding
Management shall select employee who is to be trained from the applicants.	Employee to be trained; selected from applicants in accordance with Article XVI, Section (g), Training.	Vacancy to be posted for a period of *five* calendar days, but not less than *three* production days.

COMPARISON OF ROCKY HOLLOW AND PITTSTON AGREEMENTS TO 1988 BCOA AGREEMENT—*Continued*

Rocky Hollow Agreement	1988 BCOA Agreement	Pittston Agreement
	Vacancy to be posted for a period of seven calendar days, but not less than five production days.	
Section (k) Transfer to Other Mines of Employer	Section (k) Transfer to Other Mines of Employer	Section (k) Transfer to Other Mines of Employer
Omitted from Rocky Hollow Agreement.	Same with exception noted.	Lists PCG Companies instead of employer.
Section (j) Leave of Absence	Section (l) Leave of Absence	Section (l) Leave of Absence
Employees who are elected to district or international officer shall return to a job in accordance with procedure set out in section (b) of this article—Reduction; Realignment—provided employees with greater qualifications in accordance with section (a) Definition of Seniority—are not on layoff status.	Omitted from Rocky Hollow Agreement provision for prior practice or custom for a permanent union official to return to his job.	Same as BCOA Agreement.
Section (k) Permanent and Temporary Supervisors	Section (m) Permanent and Temporary Supervisors	Section (m) Permanent and Temporary Supervisors
Employees who perform supervisory work exceeding sixty working days omitted from Rocky Hollow Agreement.		Same as BCOA Agreement.

Section (l) Shift Preference	Section (n) Shift Preference	Section (n) Shift Preference
If shift rotation not being practiced, within sole discretion of employer to assign employee with greatest qualification. BCOA Agreement: employee with greatest seniority.		Same as BCOA Agreement.
Section (m) Scope of Article		
Provides that article on seniority will not apply to any operation beyond Rocky Hollow except with respect to reciprocal agreement. (New to Rocky Hollow Agreement.)		
Article XVI—Tonnage Rates and Handloading	Article XVIII—Tonnage Rates and Handloading	Article XVIII—Tonnage Rates and Handloading
Section (a) Tonnage Rates	Section (a) Tonnage Rates	Section (a) Tonnage Rates
Tonnage rates and portal pay for handloaders omitted.		Same as BCOA Agreement.
Section (b) Checkweighman	Section (b) Checkweighman	Section (b) Checkweighman
Checkweighman or checkmeasurers shall not interfere with operation of mine.		Same as BCOA Agreement.
Section (c) Preparation and Cleaning of Coal	Section (c) Preparation and Cleaning of Coal	Section (c) Preparation and Cleaning of Coal
Same as BCOA Agreement.		Same as BCOA Agreement.

COMPARISON OF ROCKY HOLLOW AND PITTSTON AGREEMENTS TO 1988 BCOA AGREEMENT—*Continued*

Rocky Hollow Agreement	1988 BCOA Agreement	Pittston Agreement
Section (d) Delivery of Cars	Section (d) Delivery of Cars	Section (d) Delivery of Cars
Same as BCOA Agreement.		Same as BCOA Agreement.
Section (e) Explosives	Section (e) Explosives	Section (e) Explosives
Same as BCOA Agreement.		Same as BCOA Agreement.
Section (f) Bottom Coal	Section (f) Bottom Coal	Section (f) Bottom Coal
Same as BCOA Agreement.		Same as BCOA Agreement.
Section (g) Cutting Coal	Section (g) Cutting Coal	Section (g) Cutting Coal
Same as BCOA Agreement.		Same as BCOA Agreement.
Section (h) Blacksmithing	Section (h) Blacksmithing	Section (h) Blacksmithing
Same as BCOA Agreement.		Same as BCOA Agreement.
Section (i) Rockdusting	Section (i) Rockdusting	Section (i) Rockdusting
Same as BCOA Agreement.		Same as BCOA Agreement.
Section (j) Day Men Transferred	Section (j) Day Men Transferred	Section (j) Day Men Transferred
Same as BCOA Agreement.		Same as BCOA Agreement.
Article XVII—Classification	Article XIX—Classification	Article XIX—Classification
Section (a) Working in Classification	Section (a) Working in Classification	Section (a) Working in Classification
Employees cross-trained to perform different jobs.	BCOA Agreement makes provision for employee to be assigned to duties within his regular classified job.	Same as BCOA Agreement.

Section (b) Classification Requirement	Section (b) Classification Requirement	Section (b) Classification Requirement
Same as BCOA Agreement.		Same as BCOA Agreement.
Section (c) Temporary Assignments	Section (c) Temporary Assignments	Section (c) Temporary Assignments
Employer retains the right to utilize employees to perform other jobs provided employees are qualified to perform such jobs.	Omits cross-trained employees.	Same as BCOA Agreement.
Section (d) Protection Against Discrimination	Section (d) Protection Against Discrimination	Section (d) Protection Against Discrimination
Same as BCOA Agreement.		Same as BCOA Agreement.
Section (e) Compensation for Temporary Assignments	Section (e) Compensation for Temporary Assignments	Section (e) Compensation for Temporary Assignments
Specifically states that employee must work full shift when required to work outside of his classification.		Same as BCOA Agreement.
Section (f) Classification Levels		
Four Classifications: 1. Electricians 2. Underground Face Employees 3. Underground Outby Employees 4. Outside Workers (New to Rocky Hollow Agreement.)		

COMPARISON OF ROCKY HOLLOW AND PITTSTON AGREEMENTS TO 1988 BCOA AGREEMENT—*Continued*

Rocky Hollow Agreement	1988 BCOA Agreement	Pittston Agreement
Article XVIII—Health and Retirement Benefits	Article XX—Health and Retirement Benefits	Article XX—Health and Retirement Benefits
Section (a) General Purpose	Section (a) General Purpose	Section (a) General Purpose
All of BCOA Agreement section omitted with exception of first sentence making provision for pension, health and other benefits for employees.	Provides for benefits for employees covered by agreement and for employees covered under the UMWA Welfare and Retirement Fund of 1950.	Provides benefits for employees who *receive payments* under the 1974 Pension Plan and Trust. Also states that disputes arising under this article shall not be subject to settlement of disputes. Omits descriptions of 1950 Pension Plan and Trust and 1974 Pension Plan and Trust.
Section (b) UMWA 1950 and 1974 Pension Plan Contributions and Benefits Through January 31, 1988	Section (b) 1950 Plans and Trusts	Section (b) 1950 Plans and Trusts
Rocky Hollow contributed to the 1974 Pension Trust from January 12, 1986 to January 31, 1988 and to the 1950 Pension Trust from January 12, 1986 to January 31, 1988. As of February 1, 1988, Rocky Hollow's obligations to the 1974 and 1950 Pension Trusts ended.		Former employees of Pittston who retired under the 1950 Pittston Plan will continue to receive benefits from the 1950 Pension Plan and Trust. Pittston will pay to the 1950 Benefit Plan $10,000,000.

Section (c) Pension Plan Coverage Effective August 1, 1987	Section (c) 1974 Plans and Trusts	Section (c) 1974 Plans and Trusts
Rocky Hollow's obligations to 1974 Plans and Trusts terminated as of February 1, 1988. Rocky Hollow has own pension plan known as the Magnolia Plan.	1. Same in Pittston Agreement. 2. Same in Pittston Agreement. 3. (i) Insurance carrier only. Refers to "each Employer." 3. (ii) (a) and (b) Same in Pittston Agreement.	Pittston agreed to participate in the UMWA 1974 Pension Plan and Trust and the 1974 Benefit Plan and Trust. 3. (i) Provides for benefit plan through *self-insurance* or insurance carrier. Refers to *the* employer.
Section (d) Contributions by Employers	Section (d) Contribututons by Employers	Section (d) Contributions by Employers
Omitted from Rocky Hollow Agreement.		Omitted in Pittston Agreement.
Section (d) Insurance Program	Section (e) Responsibilities and Duties of Trustees	Section (e) Responsibilities and Duties of Trustees
Rocky Hollow for duration of agreement will carry a group policy of insurance for all employees and their dependents. (New to Rocky Hollow Agreement.)		Omitted in Pittston Agreement.
Section (e) General Description of Health Insurance Plan	Section (f) Audits, Reports and Notices	Section (f) Audits, Reports and Notices
New to Rocky Hollow Agreement. NOTE: Sections (e)-(j) in BCOA Agreeomitted from Rocky Hollow Agreement.		Omitted in Pittston Agreement.

COMPARISON OF ROCKY HOLLOW AND PITTSTON AGREEMENTS TO 1988 BCOA AGREEMENT—*Continued*

Rocky Hollow Agreement	1988 BCOA Agreement	Pittston Agreement
	Section (g) Administration of Trusts	Section (g) Administration of Trusts
		Omitted in Pittston Agreement.
	Section (h) Guarantee of 1950 Plans and Trusts and 1974 Plans and Trusts	Section (h) Guarantee of 1950 Plans and Trusts and 1974 Plans and Trusts
		Omitted in Pittston Agreement.
	Section (i) Withdrawal from 1974 Benefit Plan and Trust	Section (i) Withdrawal from 1974 Benefit Plan and Trust
		Omitted in Pittston Agreement.
	Section (j) Withdrawal from 1950 Benefit Plan and Trust	Section (j) Withdrawal from 1950 Benefit Plan and Trust
		Omitted in Pittston Agreement.
	General Description of the Health and Retirement Benefits	General Description of the Health and Retirement Benefits
	Pittston Agreement omits mention of 1950 Pension and Benefits Plans and Trusts.	Copies of Employee Benefit Plan available from employer. (New to Pittston Agreement.) Pittston Agreement also provides that nothing in agreement shall expand or diminish pensioners entitlement to health care benefits as existed under predecessor

	collective bargaining agreement, except for new provisions contained in the Special Health Care Cost Containment Program, the generic drug program and the Approved Provider List.
1. *Pensions for Miners Retired Under the 1950 Pension Plan*—Omitted from the Pittston Agreement.	
1A. *1950 Widows Pension*—Omitted from the Pittston Agreement.	
2. *Pensions for Miners Retired Under the 1974 Pension Plan Prior to the Effective Date of This Agreement.* Pittston Agreement lists date.	1. *Pensions for Miners Who Retired Under the 1974 Pension Plan Prior to February 1, 1988.*
(a))	(a)
(b)) No changes noted.	(b) Same as BCOA Agreement.
(c))	(c)
3. *Pensions for Miners Who Retire On or After the Effective Date of This Agreement.* Pittston Agreement lists date.	2. *Pensions for Miners Who Retire On or After February 1, 1988.*
4. *Signatory Service*	3. *Signatory Service*—Same as BCOA Agreement.
5. *Pensions for Disabled Miners.*	4. *Pensions for Disabled Miners.*
(a) Miner receives pension if accident occurs in first two years of agreement.	(a) Lists effective date of agreement, February 1, 1988. Miner receives pension if accident occurs *prior to February 1, 1990.*
(b) No changes noted.	(b) Same as BCOA Agreement.
6. *Pensions for Surviving Spouses.*	5. *Pensions for Surviving Spouses.*
(a) Effective date of this Agreement.	(a) . . . on or after February 1, 1988.

COMPARISON OF ROCKY HOLLOW AND PITTSTON AGREEMENTS TO 1988 BCOA AGREEMENT—*Continued*

Rocky Hollow Agreement	1988 BCOA Agreement	Pittston Agreement
	(b) No changes noted.	(b) Same as BCOA Agreement.
	(c) . . . on or after the effective date of this agreement.	(c) . . . on or after February 1, 1988.
	6A. *Pre-Retirement Survivor's Pension.* No changes noted.	6. *Pre-Retirement Survivor's Pension.* Same as BCOA Agreement.
	7. *Deferred Vested Pension.*	7. *Deferred Vested Pension.*
	(a) . . . effective date of this agreement. Pension calculated in accordance to number 3 above.	(a) . . . on or after February 1, 1988. Pension calculated in accordance to number 2 above.
	(b) . . . effective date of this agreement.	(b) . . . on or after February 1, 1988.
	(c) . . . effective date of this agreement.	(c) . . . prior to February 1, 1988.
	(d) No changes noted.	(d) Same as BCOA Agreement.
	8. *Life and Accidental Death and Dismemberment Benefits.*	8. *Life and Accidental Death and Dismemberment Benefits.*
	(a)) (b)) No changes noted. (c)) (d))	(a)) (b)) Same as BCOA Agreement. (c)) (d))
	9. *Pensioner's Death Benefits.*	9. *Pensioner's Death Benefits.*
	(a) Regarding 1950 Pension Plan	(a) Omitted from Pittston Agreement.
	(b) Under the 1974 Pension Plan, \$3,000 life insurance benefit paid. If no surviv-	(b) Under the 1974 Pension Plan, a life insurance benefit of *\$3,500* paid. If no

ing spouse or dependent relative, $2,500 paid.	surviving spouse or dependent relative, a life insurance benefit of *$3,000* will be paid.
10. *Health Care.* Provides that health care benefits are guaranteed during term of agreement at levels provided in 1950 Benefit Plan, 1974 Benefit Plan and Employer Benefit Plan.	10. *Health Care.* Health care benefits *provided in the Employee Benefit Plan.* Omits reference to 1950 Benefit Plan and 1974 Benefit Plan. Health care benefits provided at the level in the *Employee Benefit Plan.*
(a) Participants in 1950 Benefit Plan—Omitted in Pittston Agreement.	(a) Working miners provided health benefits through employer's benefit plan. Corresponds to (b) in BCOA Agreement.
(c) Pensioners retired under the 1974 Pension Plan provided health benefits *through* the employer from which they retired.	(b) Same as BCOA Agreement with the exception of health benefits provided *by* the employer from which they retired.
(d)) (e)) No changes noted.	(c)) (d)) Same as BCOA Agreement.
(f) Same with noted exception.	(e) Benefits for prescription drugs provided except as set forth in the Employee Benefit Plan.
(g) Same with noted exception.	(f) . . . *After satisfying applicable deductible described below* benefits for office visits to physicians, etc.

COMPARISON OF ROCKY HOLLOW AND PITTSTON AGREEMENTS TO 1988 BCOA AGREEMENT—*Continued*

Rocky Hollow Agreement	1988 BCOA Agreement	Pittston Agreement
	(h), (i), (j) No changes noted.	(g), (h), (i) Same as BCOA Agreement.
		(j) Not contained in BCOA Agreement. Provides for an Approved Provider List.
		Special Health Care Cost Containment
		New to Pittston Agreement. Provides for implementation of additional health care containment.
	Explanatory Note on Employer Provided Health Plans	*Explanatory Note on Employer Provided Health Plans*
	Same with noted exceptions.	Health care provided through *self-insurance* or insurance carriers and trustees of UMWA Health and Retirement Funds shall resolve any disputes *with respect to those provisions* which are identical to the health provisions of the 1950 Benefit Plan and Trust.
		Co-payments required under the plan in addition to satisfying deductible. If co-payment is not satisfied in first deductible year, co-payment provisions will

11. *Vision Care.*
Vision care provided through the 1950 Benefit Plan, the 1974 Benefit Plan and the Employer Benefit Plan. Coverage identical to that in 1978 Agreement.

12. *Health Care Cost Containment.*
No changes noted.

be suspended in second deductible period and then reinstated after the deductible is satisfied.

Each participant required to pay provider directly until deductible is satisfied. Claim forms will verify such payments and all payments will be applied to deductible.

Covered drug prescriptions filled at offices *on the Approved Provider List.*

11. *Vision Care.*
Omits reference to 1950 and 1974 Benefit Plans. Coverage identical to 1984 Agreement. Vision care payments subject to satisfying deductible in Special Health Care Cost Containment Program.

12. *Health Care Cost Containment.*
Same as BCOA Agreement.

13. *Most Favored Nations Clause.*
New to Pittston Agreement. Provides that if UMWA enters into an agreement which has a health care plan that's more favorable to the employer, then Pittston has the right to terminate current health care plan and implement new plan.

COMPARISON OF ROCKY HOLLOW AND PITTSTON AGREEMENTS TO 1988 BCOA AGREEMENT—*Continued*

Rocky Hollow Agreement	1988 BCOA Agreement	Pittston Agreement
Article XXA—Dental Plan	Article XXA—Dental Plan	Article XXA—Dental Plan
Omitted from Rocky Hollow Agreement.	Same with noted exception.	Same as BCOA Agreement. Refers to Pittston Coal Group Companies as employer.
Article XXB—UMWA Cash Deferred Savings Plan of 1988	Article XXB—UMWA Cash Deferred Savings Plan of 1988	Article XXB—UMWA Cash Deferred Savings Plan of 1988
Omitted from Rocky Hollow Agreement.		Omitted from Pittston Agreement.
Article XXI—Surface Mines Omitted from Rocky Hollow Agreement.	Article XXI—Surface Mines Section (a) Parking Areas	Article XXI—Surface Mines Section (a) Parking Areas
		Same as BCOA Agreement.
	Section (b) Manning of Surface Mining Equipment	Section (b) Manning of Surface Mining Equipment
		Same as BCOA Agreement.
	Section (c) Eating Place	Section (c) Eating Place
		Same as BCOA Agreement.
	Section (d) Cabs	Section (d) Cabs
		Same as BCOA Agreement.

	Section (e) Special Health and Safety Problems in Surface Mines	Section (e) Special Health and Safety Problems in Surface Mines
		Same as BCOA Agreement.
	Section (f) Toilets	Section (f) Toilets
		Same as BCOA Agreement.
	Section (g) Swing Shift	Section (g) Swing Shift
		Same as BCOA Agreement.
	Section (h) Leasing of Employees Vehicles	Section (h) Leasing of Employees Vehicles
		Same as BCOA Agreement.
	Section (i) Production and Processing of Coal at Surface Mines	Section (i) Production and Processing of Coal at Surface Mines
		Same as BCOA Agreement.
Article XIX—Miscellaneous	Article XXII—Miscellaneous	Article XXII—Miscellaneous
Section (a) Bathhouse	Section (a) Bathhouse	Section (a) Bathhouse
Same as BCOA Agreement.		Same as BCOA Agreement.
Section (b) Access Roads	Section (b) Access Roads	Section (b) Access Roads
Same as BCOA Agreement.		Same as BCOA Agreement.
Section (c) Parking Facilities	Section (c) Parking Facilities	Section (c) Parking Facilities
Same as BCOA Agreement.		Same as BCOA Agreement.

COMPARISON OF ROCKY HOLLOW AND PITTSTON AGREEMENTS TO 1988 BCOA AGREEMENT—*Continued*

Rocky Hollow Agreement	1988 BCOA Agreement	Pittston Agreement
Section (d) Bulletin Boards	Section (d) Bulletin Boards	Section (d) Bulletin Boards
Bulletin board to be used to post notices of interest to *employees.*	Bulletin board used to post notices of interest to *Union.*	Same as BCOA Agreement.
Section (e) Coke and Cleaning Plants	Section (e) Coke and Cleaning Plants	Section (e) Coke and Cleaning Plants
Omitted from Rocky Hollow Agreement.		Same as BCOA Agreement.
Section (e) Compulsory Retirement	Section (f) Compulsory Retirement	Section (f) Compulsory Retirement
Same as BCOA Agreement.		Same as BCOA Agreement.
Section (f) House Coal	Section (g) House Coal	Section (g) House Coal
Same as BCOA Agreement.		Same as BCOA Agreement.
Section (h) House Rent	Section (h) House Rent	Section (h) House Rent
Omitted from Rocky Hollow Agreement.		Same as BCOA Agreement.
Section (g) Attendance Control	Section (i) Attendance Control	Section (i) Attendance Control
Rocky Hollow absenteeism policy more complicated than BCOA policy.		Same as BCOA Agreement.
Section (j) Memorial Periods	Section (j) Memorial Periods	Section (j) Memorial Periods
Omitted from Rocky Hollow Agreement.		Same as BCOA Agreement.

Section (h) Closing Following Fatal Accident	Section (k) Closing Following Fatal Accident	Section (k) Closing Following Fatal Accident
Eliminates reference to Memorial Period.		Same as BCOA Agreement.
Section (i) New Machinery	Section (l) New Machinery	Section (l) New Machinery
No notice given to Union for installing new equipment. Employer to decide manning requirements.		Same as BCOA Agreement.
Section (j) Pay Day	Section (m) Pay Day	Section (m) Pay Day
Same as BCOA Agreement.		Same as BCOA Agreement.
Section (k) Lunches	Section (n) Lunches	Section (n) Lunches
Employee works *four* or more hours in addition to regular shift and if called back gets lunch at employer's expense.	Employee works *two* or more hours.	Same as BCOA Agreement.
Section (l) Portals	Section (o) Portals	Section (o) Portals
Employer has the right to change portals without union approval.		Same as BCOA Agreement.
Section (m) Tools	Section (p) Tools	Section (p) Tools
Same as BCOA Agreement.		Same as BCOA Agreement.
Section (q) Tramming	Section (q) Tramming	Section (q) Tramming
Omitted from Rocky Hollow Agreement.		Same as BCOA Agreement.

COMPARISON OF ROCKY HOLLOW AND PITTSTON AGREEMENTS TO 1988 BCOA AGREEMENT—*Continued*

Rocky Hollow Agreement	1988 BCOA Agreement	Pittston Agreement
Section (n) Local Union Meeting Place	Section (r) Local Union Meeting Place	Section (r) Local Union Meeting Place
Bathhouse used as a meeting place if it doesn't interfere with intended use of the *mine.*	Bathhouse used as a meeting place if it doesn't interfere with intended use of *facilities.*	Same as BCOA Agreement.
Section (o) Bonus Plans	Section (s) Bonus Plans	Section (s) Bonus Plans
1. *Procedures:* The employer, within sole discretion, may establish, revise or terminate a bonus plan.	1. Procedures, subsections (a)-(f)—No changes noted in Pittston Agreement.	1. Procedures, subsections (a)-(f)—Same as BCOA Agreement.
1. Subsection (a)—Same as BCOA Agreement.		
1. Subsection (b)—Employer provides description of bonus plan thirty days before commencement to Mine Committee to discuss plan.	1. Subsection (b)—Provides sixty-day period to discuss bonus plan and Mine Committee has the right to initiate a grievance if bonus plan doesn't meet requirements.	
1. Subsection (c)—Employer shall have sole discretion to initiate, terminate or modify bonus plans and shall give thirty days notice of modification or termination.	1. Subsection (c)—(Corresponds to subsection (b) in Rocky Hollow Agreement)—Employer has forty-five days to distribute a description of plan to employees and Mine Committee.	

2. *Conditions:* Omitted from Rocky Hollow Agreement provision for bonus plan lessening safety standards. 2. Subsection (a)—Omitted from Rocky Hollow Agreement. 2. Subsection (b)—Same as BCOA Agreement. 2. Subsection (c) and (d)—Omitted from Rocky Hollow Agreement.	1. Subsection (d)—(Corresponds to subsection (c) in Rocky Hollow Agreement)—Employer must explain any revisions to Mine Communication Committee after giving seven days notice of such change. (Section (e)-(f) of BCOA Agreement omitted from Rocky Hollow Agreement)	2. Conditions, subsections (a)-(d)—Same as BCOA Agreement.
3. *Resolution of Disputes:* Omits reference to bonus plan that lessens safety standards.	3. Resolution of Disputes—No changes noted in Pittston Agreement.	3. Resolution of Disputes—Same as BCOA Agreement.
Article XX—Settlement of Disputes	Article XXIII—Settlement of Disputes	Article XXIII—Settlement of Disputes
Section (a) Mine Committee	Section (a) Mine Committee	Section (a) Mine Committee
Mine Committee to consist of no more than three *active* employees.	Mine Committee to consist of three but no more than five employees.	Same as BCOA Agreement.
Section (b) Arbitrators	Section (b) Arbitrators	Section (b) Arbitrators
Numbers 1 and 2 combined into number 1 in Rocky Hollow Agreement. BCOA Agreement allows ninety days to select impartial arbitrators. Rocky Hollow Agreement sets thirty days.	1. President of the UMWA International Union and president of BCOA shall jointly establish a panel of arbitrators. Does not provide for number of arbitrators to serve on panel.	1. *Employer* and International Union to establish a panel of arbitrators. Panel to consist of ten arbitrators. Arbitrator to serve for a term of eighteen months. At the end of term, each arbitrator re-

Rocky Hollow Agreement	1988 BCOA Agreement	Pittston Agreement
Rocky Hollow panel to have six arbitrators. BCOA Agreement provides for no specific number for each UMWA district. In the Rocky Hollow Agreement, provision for Step 4 to provide final and binding arbitration of the dispute omitted. Number 4 in Rocky Hollow Agreement omits that arbitrator's failure to render an expeditious decision may be grounds for removal. Also eliminates ten-day limit after union and employer fail to agree on settlement of dispute for its referral to an arbitrator. Eliminates consolidation of cases. Provides for post hearing briefs to be filed by either party and to be part of record. BCOA Agreement does not permit briefs unless arbitrator determines their necessity.	2. Same with noted exception in Pittston Agreement. 3. No changes noted. 4. No changes noted.	viewed and reappointment considered. 2. . . . *the* employer and the UMWA district to select a district arbitrator. 3. Same as BCOA Agreement. 4. Same as BCOA Agreement.
Section (c) Grievance Procedure	Section (c) Grievance Procedure	Section (c) Grievance Procedure
During first six months of Rocky Hollow Agreement, only disputes involving discipline or discharge arbitrated.		

(1) Same as BCOA Agreement. (2) Eliminates provision that Mine Committee consist of more than three members. Also, complaints submitted on Standard Grievance Form to be agreed upon within thirty days of signing of Agreement. (3) Provides for transcript to be taken.	(1) No changes noted. (2) Complaint submitted on BCOA-UMWA Standard Grievance Form. (3) No changes noted in Pittston Agreement.	(1) Same as BCOA Agreement. (2) Complaint submitted on *Employer*-UMWA Standard Grievance Form. (3) Same as BCOA Agreement.
(4) Where district representative and employer fail to reach agreement, the matter then referred to arbitrator jointly agreed to by both parties. There is no ten calendar-day time limit for matter to be referred to arbitrator as provided for in BCOA Agreement.	(4) When there is a failure to reach agreement, the matter shall be referred to a district arbitrator. Cases assigned to district arbitrators in rotation. Hearings take place at location agreed upon by the parties. If the parties cannot agree upon a hearing place, the umpire shall select the place.	(4) When there is a failure to reach agreement, the matter shall be referred to *panel* arbitrator. Cases assigned to *panel* arbitrators utilizing a lottery system. Hearings take place at location agreed upon by *representatives of the union and employer.* If *these representatives* cannot agree upon a hearing place, the umpire shall select the place.
Omits section in BCOA Agreement relating to grievances pending same operation of employer being assigned to single arbitrator and heard on the same day. Post hearing briefs permitted. In BCOA Agreement, post hearing briefs shall not be permitted except in cases where the arbitrator determines their necessity.	The arbitrator advises the parties if he can't make a decision within thirty days.	The arbitrator advises the *union and employer* if he can't make a decision within thirty days.

COMPARISON OF ROCKY HOLLOW AND PITTSTON AGREEMENTS TO 1988 BCOA AGREEMENT—*Continued*

Rocky Hollow Agreement	1988 BCOA Agreement	Pittston Agreement
Section (d) Ten-Day Limitation	Section (d) Ten-Day Limitation	Section (d) Ten-Day Limitation
Grievances not filed within ten days of the occurrence of the event upon which grievance is based upon shall be denied as untimely and not processed further.	Grievances not filed within ten days of the time when employee reasonably should have known it.	Same as BCOA Agreement.
Section (e) Earnest Effort to Resolve Disputes	Section (e) Earnest Effort to Resolve Disputes	Section (e) Earnest Effort to Resolve Disputes
Same as BCOA Agreement.	*Company representatives* shall disclose facts relied upon by *company*.	Same as BCOA Agreement with the exception that company representative changed to *employer* representatives and company changed to *employer*.
Section (f) Employee's Right to Presence of Member of Mine Committee	Section (f) Employee's Right to Presence of Member of Mine Committee	Section (f) Employee's Right to Presence of Member of Mine Committee
Same as BCOA Agreement.		Same as BCOA Agreement.
Section (g) Right of Grievant to be Present	Section (g) Right of Grievant to be Present	Section (g) Right of Grievant to be Present
Grievant attends such grievance proceedings without pay and at own expense.		Same as BCOA Agreement.

Section (h) Finality of Decision or Settlement	Section (h) Finality of Decision or Settlement	Section (h) Finality of Decision or Settlement
Same as BCOA Agreement.		Same as BCOA Agreement.
Section (i) Waiver of Time Limits	Section (i) Exclusion of Legal Counsel	Section (i) Exclusion of Counsel
Same as BCOA Agreement. Corresponds to Section (j) in BCOA Agreement.		Same as BCOA Agreement.
Section (j) Exclusion of Legal Counsel	Section (j) Waiver of Time Limits	Section (j) Waiver of Time Limits
Same as BCOA Agreement. Corresponds to Section (i) in BCOA Agreement.	The "parties" may waive time limits.	*Representatives of the union and employer* may waive time limits.
Section (k) Limitation on Dispute Resolution	Section (k) Prior Agreement	Section (k) Prior Agreement
New section added to Rocky Hollow Agreement. Articles relating to settlement of disputes terminate with expiration of agreement.	This section omitted from Rocky Hollow Agreement.	Same as BCOA Agreement.
Article XXI—Discharge Procedure	Article XXIV—Discharge Procedure	Article XXIV—Discharge Procedure
Section (a) General	Section (a) Just Cause Required	Section (a) Just Cause Required
Gives employer the right to discharge, suspend or otherwise discipline employees for just cause.		Same as BCOA Agreement.

COMPARISON OF ROCKY HOLLOW AND PITTSTON AGREEMENTS TO 1988 BCOA AGREEMENT—*Continued*

Rocky Hollow Agreement	1988 BCOA Agreement	Pittston Agreement
Section (b) Procedure	Section (b) Procedure	Section (b) Procedure
Applies only during term of Agreement only.		Same as BCOA Agreement.
Section (c) Suspension	Section (c) Suspension	Section (c) Suspension
Permits employee who is suspended with intent to discharge to file a grievance and proceed to Step 3 of the grievance procedure.		Same as BCOA Agreement.
Section (d) Immediate Arbitration	Section (d) Imediate Arbitration	Section (d) Immediate Arbitration
Provides that if arbitrator determines just cause for discharge, the discharge becomes effective immediately instead of date of decision as provided for in the BCOA Agreement.		Same as BCOA Agreement.
Section (e) Regular Arbitration	Section (e) Regular Arbitration	Section (e) Regular Arbitration
Same as BCOA Agreement.		Same as BCOA Agreement.
Section (f) Compensation for Lost Earnings	Section (f) Compensation for Lost Earnings	Section (f) Compensation for Lost Earnings
Employee reinstated subject to facts and paid straight and premium time rates less any income from other source.		Same as BCOA Agreement.

Section (g) Discharge as a Result of Incident Occurring After Expiration of This Agreement		
If employee discharged for any acts while engaged in a strike or work stoppage after the expiration of this Agreement, he shall have no recourse through the arbitration procedure. (New to Rocky Hollow Agreement.)		
Article XXII—No Strike/No Lockout Clause		
New to Rocky Hollow Agreement. Not contained in the BCOA Agreement. No strikes, slowdowns, walkouts, etc. permitted during the term of Rocky Hollow Agreement.		
Article XXIII—Discrimination Prohibited	Article XXV—Discrimination Prohibited	Article XXV—Discrimination Prohibited
Same as BCOA Agreement.		Same as BCOA Agreement.
Article XXIV—District Agreements	Article XXVI—District Agreements	Article XXVI—District Agreements
Section (a) New Districts	Section (a) New Districts	Section (a) New Districts)
Same as BCOA Agreement.		Same as BCOA Agreement.

COMPARISON OF ROCKY HOLLOW AND PITTSTON AGREEMENTS TO 1988 BCOA AGREEMENT—*Continued*

Rocky Hollow Agreement	1988 BCOA Agreement	Pittston Agreement
Section (b) Prior Practice and Custom	Section (b) Prior Practice and Custom	Section (b) Prior Practice and Custom
Reference to all prior practice and custom, except where abolished by mutual agreement and not in conflict with agreement, omitted in Rocky Hollow Agreement.	Within six months employer and all UMWA District presidents provide the International Union and BCOA with copies of agreements.	The employer and all of the UMWA District presidents shall *exchange* copies of all District Agreements.
Section (c) Protective Wage Clause	Section (c) Protective Wage Clause	Section (c) Protective Wage Clause
Omitted in Rocky Hollow Agreement.		Same as BCOA Agreement.
Section (d) Approval of District Agreements	Section (d) Approval of District Agreements	Section (d) Approval of District Agreements
Omitted in Rocky Hollow Agreement.		Same as BCOA Agreement.
Article XXV—Maintain Integrity of Contract and Resort to Courts	Article XXVII—Maintain Integrity of Contract and Resort to Courts	Article XXVII—Maintain Integrity of Contract and Resort to Courts
For term of agreement only. No successorship provision for union to seek judicial relief in cases involved in successorship.		Same as BCOA Agreement.
Article XXVI—Severability Clause	Article XXVIII—Severability Clause	Article XXVIII—Severability Clause
Section (a) General Rule	Section (a) General Rule	Section (a) General Rule
Same as BCOA Agreement.		Same as BCOA Agreement.

Section (b) Exception	Section (b) Exception	Section (b) Exception
Adds that in the event the employer is unable to pass through costs to customers possibly resulting in loss of job opportunities for employees, then either party with sixty days notice may terminate agreement.		Same as BCOA Agreement.
Article XXVII—Zipper Clause		
Gives both parties option to change agreements by giving written notice and that upon expiration of agreement, all terms and conditions of employment will completely terminate. (New to Rocky Hollow Agreement.)		
Article XXVIII—Ratification and Termination of This Agreement	Article XXIX—Ratification and Termination of This Agreement	Article XXIX—Ratification and Termination of This Agreement
The Rocky Hollow Agreement is a two and one-half year agreement commencing on October 30, 1988 and terminating on April 30, 1991. In the event no successor agreement is reached by the parties as of April 30, 1991, the terms of the Rocky Hollow Agreement will extend to October 31, 1991.	The BCOA Agreement is a five-year Agreement commencing on February 1, 1988 and terminating February 1, 1993. It also provides that the Union may reopen the agreement for renegotiating wage rates and pension benefits upon the third and fourth anniversaries of the agreement.	The Pittston Agreement is a four-year agreement which became effective upon ratification by the Pittston miners on February 19, 1990 and terminating June 30, 1994.

COMPARISON OF ROCKY HOLLOW AND PITTSTON AGREEMENTS TO 1988 BCOA AGREEMENT—*Continued*

Rocky Hollow Agreement	1988 BCOA Agreement	Pittston Agreement
Appendix Part A and B Omitted from Rocky Hollow Agreement.	Appendix Part A and Part B	Appendix Part A and Part B
Index	Index	*New to Pittston Agreement:*
		Appendix C—Flextime Scheduling
		Appendix D—Memorandum of Understanding Regarding Profit-Sharing Plan
		Appendix E—Memorandum of Understanding Regarding Wage Increases in 1993
		Appendix F—Memorandum of Understanding Regarding Ratification Bonus
		Appendix G—Memorandum of Understanding Regarding Continuous Mining Machine Helpers
		Appendix H—Memorandum of Understanding Regarding Utilization of Memorial Periods

APPENDIX IV

Quarterly Update (Period Ending March 31, 1990) of the UMW Selective Strike Fund

United Mine Workers of America

JOHN J. BANOVIC
INTERNATIONAL
SECRETARY-TREASURER

TELEPHONE
(202) 842-7260

UNITED MINE WORKERS' BUILDING
900 FIFTEENTH STREET, N. W.
Washington, D.C.
20005

April 20, 1990

Dear Brothers and Sisters:

The following is the quarterly update on the UMWA Selective Strike Fund for the period ending March 31, 1990.

The Fund decreased over $7 million this quarter due to the (.5%) assessment rate and the number of members who are currently receiving benefits. The total number of members receiving Selective Strike Benefits totalled 2,105 as of the end of March, 1990.

Administrative cost related to the Selective Strike Fund exceeded $250,000 for the first quarter of 1990.

Balance December 31, 1989	$ 71,261,275
Assessment Income	3,135,865
Investment Income	976,933
Strike Benefits Paid	(8,250,335)
Insurance Benefits Paid	(3,000,000)
Bank Charges	(3,288)
March 31, 1990	$ 64,120,450

The current financial update of the benefits paid to each local since the inception of the Selective Strike Fund is enclosed.

Please contact my office should you have any questions regarding this report.

Fraternally,

John J. Banovic

JJB/law

cc: All Local Union Presidents
All Local Union Financial Secretaries
All Local Union Recording Secretaries
All District Presidents
All District Secretary-Treasurers
All IEB Members
All International Auditors & Tellers

UMWA INTERNATIONAL
SSF BENEFITS PAYMENTS
SINCE INCEPTION 1984 THROUGH MARCH, 1990

District	# of People	Dollar Amount
2	413	$ 7,665,172
4	241	$ 2,929,990
5	621	$ 10,126,691
6	41	$ 50,000
15	1,464	$ 10,653,837
17	1,583	$ 18,621,640
20	818	$ 6,179,010
22	1,644	$ 1,986,800
23	116	$ 1,308,949
25	841	$ 6,233,615
28	1,758	$ 18,564,615
29	1,201	$ 10,872,915
30	2,196	$ 34,188,847
31	45	$ 91,650
Total	12,982	$129,473,731

APPENDIX V

Decision of Circuit Judge Donald McGlothlin, Jr. August 22, 1990 Russell County, Virginia

Commonwealth of Virginia

CIRCUIT COURT JUDGES

NICHOLAS E. PERSIN
P. O. Box 606
Grundy, Virginia 24614
(703) 935-2451

TWENTY-NINTH JUDICIAL CIRCUIT
COUNTIES OF BUCHANAN, DICKENSON, RUSSELL AND TAZEWELL

DONALD A. McGLOTHLIN JR.
Russell County Courthouse
Lebanon, Virginia 24266
(703) 889-8049 (Lebanon)
(703) 926-4856 (Clintwood)

DONALD R. MULLINS
Tazewell County Courthouse
Tazewell, Virginia 24651
(703) 988-5998

August 22, 1990

Stephen M. Hodges, Esq.
P.O. Box 2288
Abingdon, VA 24210

Wade W. Massie, Esq.
P.O. Box 2288
Abingdon, VA 24210

James J. Vergara, Jr., Esq.
100 Main Street Plaza
Hopewell, VA 23806

William O. Shults, Esq.
U.M.W.A.
International Union
900 15th Street N.W.
Washington, D.C. 20005

James M. Haviland, Esq.
124 Capitol Street
Charleston, West VA 25301

Re: Clinchfield Coal Company, et al v.
International Union, United Mine Workers,
of America, et al
Chancery No. 12486

Gentlemen:

This matter is before the Court upon the defendants motion to set aside the December 15, 1989, orders previously entered by this Court, upon the parties' joint motion for "Order Re Dismissal," and upon the several representations of counsel and memoranda filed in support of the parties' positions herein. The Court should note also that two memoranda have been filed by the Center on National Labor Policy Inc., as *amicus curiae,* in opposition to the plantiffs' and defendants' joint motion.

On December 15, 1989, this Court entered three orders, one liquidating prospective fines under its sixth contempt order, and two others adjudicating the defendant in contempt and liquidating portions of the prospective fines which had been previously announced on May 16 and June 2, 1989. On January 5, 1990, at 11:30 a.m., the defendants filed a motion to set aside these three orders as being contrary to the law and the evidence along with a motion to postpone the hearing on the motion to set aside. At this time the parties announced a tentative settlement of the strike "underlying this action." The Court entered its order that day temporarily suspending the three orders on December 15, in order to provide the defendant more time to fully present its arguments and the Court more time to consider the motion to set aside and any other motions the parties might file thereafter.

After having considered the argument of counsel and the authorities submitted, it is the opinion of this Court that the defendants' motion to set aside the orders entered December 15, 1989, must be denied.

The evidence presented by the plaintiffs as to each of the allegations of contemptuous behavior proves without question that the International U.M.W.A. was the author of these actions. The fact that some of the evidence as to the defendants' complicity was circumstantial makes it no less competent or convincing.

Nor can the defendants' protestations that these proceedings were criminal contempt hearings change the fact that they were civil in nature. The Court early on announced its purpose in imposing prospective civil fines the payment of which would only be required if it were shown the defendants disobeyed the Court's orders. That purpose was to compel compliance with the Court's orders which were entered to protect the rights of the plaintiffs and the public. The fines were conditional and it was within the defendants' sole power to avoid payment of the fines.

The Supreme Court of the United States in *United States v. United Mine Workers of America,* 330 U.S. 258, 304-305, 67 S.CT. 677, 701

(1947) has specifically sanctioned the imposition of monetary fines payable to the Court to compel a recalcitrant defendant (the same defendant as in this case) to discontinue a strike it had called. The United States Court of Appeals for the Second Circuit has likewise found that the imposition of fines *"in terrorem"* is authorized as a means of securing future compliance with a decree. *Sunbeam Corp. v. Golden Rule Appliance Co., Inc.,* 252 Fed. 467, (1958). In that case the defendant was ordered to pay a competitor a $2,500.00 fine for every future sale of the competitor's product in violation of a consent decree. The fine there, as here, was imposed only conditionally and depended upon the contemnor's future conduct. The contempt sanctions imposed by this Court were civil in nature, not criminal. Thus, defendant had no right to trial by a jury or to a public prosecutor.

Defendants complained of the Court's appointment of counsel for plaintiffs as Special Commissioners to collect the fines imposed. This issue is now moot as Mssrs. Hodges and Massie have been relieved of those duties.

The remaining grounds assigned by defendants in its motion to set aside are without merit and the Court overrules the motion. The Court would comment, though, that the interpretation of the evidence with regard to the existence of "roving pickets" and the understanding of the bases of the Court's findings argued by defendants' counsel belie their intelligence and are tributes to their inventiveness. This Court in its earlier orders specified the sites where picketing would be authorized along with the number of pickets allowed at each site. The evidence presented at all of the hearings showed a constant pattern of pickets locating themselves at various unauthorized places from which they would move to another picket site when police or others attempted to investigate incidents. The evidence of moving or roving picketing was overwhelming.

The parties filed a joint motion for "Order Re Dismissal" on January 24, 1990, when they again represented to the Court that they had come upon a tentative settlement of the strike, and asked the Court to rule immediately on this motion even though the proposed agreement had to be submitted to the U.M.W.A. members for a ratification vote—something which had not even been scheduled at that time. The Court conducted two hearings during which it was shown a written agreement entered into by the parties purportedly resolving the many cases in litigation spawned by the strike. At the parties' request the Court viewed *in camera* a supplemental agreement between them containing portions of their pact which the International Union desired not be made public. Certain "submissions" were made by the defendant in which it proposed to have its membership perform community service work in order to purge itself of contempt. On February 12, 1990, the Court ruled from the bench that it would not, upon the evidence, the representations and the arguments presented to that date vacate its orders imposing the civil fines merely because the parties agreed it should do so. Thereafter, an additional "submission" was filed by the defendant, increasing the number of hours of community service proposed, together with several additional memoranda by Clinchfield, the U.M.W.A. and the Center on National Labor Policy, Inc. (as *amicus curiae*). Subsequently there was also proposed to the Court a hearing at which the top leadership and management of the parties would appear to discuss various issues with the Court. The Court was of course available for such a hearing, but none was ever scheduled even though several weeks passed after the proposal was made. It appearing that nothing further will be forthcoming on these issues, the Court considers them ripe for adjudication.

The parties have requested dismissal of this lawsuit and dissolution of the injunctions entered. There is no question that upon reaching a settlement of their dispute these litigants are entitled to have such requests granted. The Court will, therefore, enter an order dismissing plaintiffs' Bill of Complaint for injunctive relief and dissolving all injunctions insofar as they grant relief to the plaintiffs.

The parties have also requested vacation of all orders imposing fines for contempt, including not only those entered December 15, 1989, but also judgments entered more than twenty-one days prior to the filing of the motion. Virginia Supreme Court Rule 1:1 provides that "All final judgments, orders and decrees, . . . shall remain under the control of that trial court for twenty-one days after the dates of entry, *and no longer.* (Emphasis supplied.) Defendants have noted appeals of the order entered before December 15, 1989, and they are now before the Court of Appeals of Virginia. Having taken the position that these orders are final and therefore ripe for appeal (a position with which this Court agrees), the defendants cannot now argue to the contrary. These orders are beyond the reach of this Court and shall remain undisturbed.

Turning to the December 15 orders now, the Court is told that because this is a civil suit, the litigants' are entitled, upon announcing a settlement of their disputes, to have the contempt proceedings dismissed and all fines previously liquidated vacated. The Court is cognizant of the principles by law upon which the argument is founded but disagrees with the contention that they require in this case the action requested by the Court.

The underlying action upon which these civil contempt proceedings are dependant is, of course, plaintiffs' suit for injunctive relief, the purpose of which was to obtain the preventative power of a Court of Equity 1) to protect the company from the power of a large labor union unlawfully used against it, and 2) to protect its employees, servants, contractors, their families and members of the general public from defendants' unlawful acts. Clinchfield sought government intervention to prevent the obstruction of private and public rights of ways; the intimidation and coercion of any persons entering or leaving plaintiffs' worksites; the threatening and assaulting of persons; the throwing of rocks and other missiles at vehicles or persons; the placing of devices designed to puncture tires of automotive traffic on private and public roads; the obstruction of the vision of those operating motor vehicles; the following or trailing of company employees and their families, and so on. The Court granted much of the relief requested finding defendants were interfering with the rights of the plaintiffs *and* those of the general public.

Moreover, with the passage of time defendants' strategy for conducting the strike shifted from acts affecting primarily the company to acts affecting both those members of the public associated with the company and those who had no connection with any of the litigants. The focus and loci of defendants' unlawful conduct shifted from company property and facilities to the public highways and private homes and businesses. So, as the strike proceeded, the protection sought from and granted by the Court was more and more for the general public. Concomitant action by the executive branch of this Commonwealth's government brought scores of Virginia State Police officers, Department of Transportation workers, etc., and millions of dollars worth of equipment to the task of protecting the rights of not only the litigants, but the general public as well. Thus, when the Court found it necessary to liquidate the prospective fines imposed, and when defendant strenuously objected to awarding them to Clinchfield, the bulk of the fines were made payable to the Commonwealth and the two counties most heavily affected by the unlawful activity. From its institution, and more and more as it proceeded, this case has involved the protection of the rights of the general public in addition to those of the plaintiffs. In *Gompers v. Buck Stove & Range Co.* 221 U.S. 418 (1911), the United States Supreme Court employed an analysis of the purpose of a suit and the relief awarded in determining whether litigants are entitled to a dismissal of contempt proceedings. Although in that case the Court ruled in favor of the requested dismissal, analysis of the purpose of this suit and the nature of the relief awarded shows there are substantial differences in the facts there and here, differences so significant as to compel an opposite result.

Additionally, the motion for "Order Re Dismissal" was not made until January 24, 1990, 19 days after the Court agreed to suspend the operation of its December 15, 1990 orders concerning contempt. Even at that time their request was not to dismiss the cause, but to enter an order stating the Court's *intention* to dismiss it upon the ratification of the tentative labor contract and cessation of the strike. The Court's rulings finding the defendant guilty of several counts of contempt, liquidating numerous fines and apportioning them among the plaintiff, Dickenson and Russell Counties and the Commonwealth were announced from the bench December 8, 1989, and reduced to written form by the orders entered December 15, 1989. The effects of these rulings were suspended by order entered January 5, 1990, to allow the Court time to consider defendants motion to set aside. The request for vacation of the contempt fines came too late.

What is more, the parties announced only *tentative settlement* of of their dispute. Even the language of the proposed order was conditional, only to be given effect should the proposed labor contract be ratified by the U.M.W.A. membership. Defendants even argued that the Court's entry of the proposed order was a condition prerequisite to the submission of the contract proposal to its membership. Although the Court is aware that the proposal was indeed ratified by the unions' membership in February, 1990, and the Court assumes that the strike and picketing against plaintiffs has indeed ceased, the Court feels strongly that the parties attempted by these motions to set aside and to dismiss and vacate to manipulate the Court's decision making process and the orderly disposition of these matters. Basically, the parties attempted to extort the desired ruling from the Court by making it appear final settlement of this bitter, violent labor dispute was contingent upon that favorable ruling. The parties did not act in good faith.

Although the law favors the resolution of disputes by compromise and settlement rather than by litigation, *Bangor Punt Operations, Inc, v. Atlantic Leasing Ltd.,* 215 Va. 180, 207 S.E. 2d 858, the reason for this rule is instructive. Settlement is less expensive and less time consuming. It saves time, effort and expense of the parties, the attorneys and the courts Compromise and settlement is said to be conducive to more amicable relations between the parties. (*15 Am. Jur. 2d Compromise and Settlement,* Section 6). Here, no time can be saved nor expense avoided. The parties' time, effort and money have already been expended; nor will the attorneys' or Court's time be saved—it has already been consumed; nor will future relations of the parties be affected when both parties have joined in the motion to dismiss/vacate and done their utmost to obtain a favorable ruling.

Here, the parties rested their cases, submitted the issues to the Court, and the Court ruled and entered its orders. In effect the original parties came upon a settlement of their differences after judgment. Had the rulings been such that Clinchfield was the sole beneficiary of the judgments of the Court, it would be simple enough for the parties to effect their object. The judgments, however, made the Commonwealth and two of its political subdivisions recipients of portions of the liquidated fines. Due to the nature of defendants' unlawful conduct, the nature of the relief requested by Clinchfield, and the protection the Court granted, the public, the citizens of this Commonwealth, were necessarily intimately affected and were the intended beneficiaries of the suit and the relief granted. There is a decisive difference between such a case and one involving only the litigants or one in which the Court-ordered protection is focused solely on the litigants. The Supreme Court of the United States has recognized that the payment of fines to the Court is proper remedial relief when payment can be avoided by compliance with the Court's order. *Hicks v. Feioks,* 485 U.S. 624, 99 L. Ed. 2d 721 (1988). Certainly the same is true of fines the Court makes payable to other branches of the government. Where, as here, the public's welfare is so intimately involved and the Court has granted civil contempt relief payable in effect to the public, and where judgment has been announced and entered, the public's interest must be considered. There can be no "settlement" without the consent of at least all those whom the relief granted is intended to benefit. No such consent is present in this case. Neither the Commonwealth, nor the two counties, nor the court has agreed to the vacation of these fines.

The Court must also take into consideration the fact that the defendants have made no meaningful effort to purge their contempt. The offer to have its members perform a paltry 10,000 or even 20,000 hours of community service to atone for the repeated, massive, violent violations of this Court's orders is an affront to our system of law and to the Court. Likewise, if ever a party has come to the bar seeking equity with unclean hands, the defendants in the case have. Not only have they violated the injunctions put down to protect the plaintiffs and the public, but they have refused to abide by the Court's judgments, e.g.: directing payment of fines and service of the orders upon its membership. The International, U.M.W.A. remains defiant and deserves no relief.

This Court's judgments will be given effect to the extent that consent to vacate them is lacking. As the plaintiffs have indicated their consent, and since they can easily have any judgment in their favor marked "satisfied" by the Clerk, so much of the fines as were directed payable to Clinchfield shall be vacated. The remainder shall be paid by the defendants with interest from December 15, 1989. The suspension of those portions of all orders adjudicating contempt, liquidating fines against the defendants, and directing the method and time of payment of the fines shall be terminated and they shall remain intact and in effect. All fines liquidated by the orders entered December 15, 1989, shall be payable to the recipients through the Clerk of this Court no later than 10 days from the entry of the order commemorating the rulings herein.

The Court will appoint John L. Bagwell, Esquire, of Grundy, Virginia, to act in the place and stead of the Commonwealth's Attorneys of Russell and Dickenson Counties, who have both asked to be disqualified in all these cases, and to act as Special Commissioner in Chancery for the purpose of collecting any unpaid fines due and payable to those political subdivisions and the Commonwealth at a fee to be approved by the Court. The Court shall further direct Mr. Bagwell to take all actions necessary to immediately begin collection of any fines remaining unpaid after the date specified above and to report to the Court, his efforts and the results of those efforts at collection.

In its comments from the bench to response to the motions disposed of by the rulings in this opinion, the Court expressed its concern over the need to establish and protect the rule of law and the authority and power of courts to enforce the law. The conduct of the defendants throughout the history of this litigation has certainly given rise to grave concerns over whether they will be governed by the law and the institutions created by to administer the law, or whether they will be permitted to operate outside the rules society establishes for the conduct of affairs amongst its members. Because the judgments heretofore entered are civil in nature, and because the contempt proceedings previously had were for the purpose of persuading the defendants to stop violating the rights of plaintiffs and the citizens of these communities, and because there is evidence before the Court that the defendants have violated several of this Court's orders, and because it is imperative that if the defendants have knowingly violated these orders, they must be made to realize the consequences thereof, the Court is of the opinion that criminal contempt proceedings must be instituted to determine whether the defendants or its members have been guilty of knowingly violating these orders.

Mr. Hodges is directed to prepare at the very earliest date practicable an appropriate order in accordance with the rulings herein for endorsement by counsel and entry by the Court.

Very truly yours,

Donald A. McGlothlin, Jr., Judge

DAMjr/dpr

Bibliography

Alinsky, Saul D. *John L. Lewis: An Unauthorized Biography.* Vintage Books, New York, 1970, pp. 15-17; 21-22, 188-91.

Associated Press. "Trumka urges coalfield visits; union is fined $4.5 million," *Williamson Daily News,* July 27, 1989.

Associated Press. "Outcome of strike may be key to coal future; Pittston Co. wants out of health plan," *Williamson Daily News,* August 28, 1989.

Associated Press. "Judge says he'll delay more fines vs. union," *Williamson Daily News,* October 31, 1989.

Balow, Jim. "UMW executive board votes to rejoin AFL-CIO," *Charleston Gazette,* October 5, 1989.

Bragg, Steve. "Wife of convicted sniper: 'Our husbands gave up their lives. . . ,'" *Williamson Daily News,* June 7, 1989.

Brown, Warren. "UMW Bank Sale: Poetic Justice; Union Embraces The Establishment," *Washington Post,* March 6, 1985.

Carter, Chuck. "Solidarity was the key word at UMWA rally," *Logan Banner,* June 12, 1989.

Charleston Gazette. "UMW Leader Blizzard Dies," August 1, 1958.

Charleston Gazette. "Compare the Pittston offer to the A. T. Massey Companies' Agreement and the Benjamin Coal Company's," May 11, 1989.

Deimel, Robert William. *The Public Relations Activity of the United Mine Workers of America Between 1946 and 1950; A Contribution to the History of Public Relations.* Thesis submitted to the Graduate School of West Virginia University, Morgantown, West Virginia, 1971, pp. 204-33.

Dubofsky, Melvin and Van Tine, Warren. *John L. Lewis: A Biography.* Quadrangle/New York Times Book Co., Inc., New York, 1977, p. 222.

Gaither, Bill. "UMW official shot, wounded in confrontation," *Sunday Gazette-Mail,* August 20, 1989.

Greiner, Ruth Jones, C.S.R., R.P.P. Court Reporter's Transcribed Record of Judge Donald McGlothlin, Jr. Order June 2, 1989, Russell County Circuit Court, Lebanon, Virginia, pp. 6-8.

Harris, V. B. *Kanawha's Black Gold and the Miner's Rebellions.* Unpublished manuscript, Chapter Nine, Arnold Miller, UMW President, 1972-1979.

Harris, V. B. *Kanawha's Black Gold and the Miner's Rebellions.* Unpublished manuscript, Chapter Ten, Sam Church, UMW President, 1979-1982.

International Executive Board, United Mine Workers of America. *John L. Lewis and the International Union, United Mine Workers of America: The Story From 1917 to 1952.* Published by authority of the International Executive Board of the United Mine Workers of America, 1952, pp. 71-72; 90-91; 103; 134-35; 147; 149; 157-59; 161; 165.

Lovegrove, Richard. "AFL-CIO president arrested in show of support for miners," *Charleston Gazette,* August 24, 1989.

National Coal Association. *1983 Facts About Coal.* National Coal Association Public and Media Affairs Group, Washington, D.C., 1983, p. 18.

National Coal Association. *Facts About Coal 1988.* National Coal Association Public and Constituent Relations, Washington, D.C., 1987, p. 30.

NLRB v. Jones & Laughlin Steel Corporation, 301 U. S. 1 (1937).

Nussel, Philip. "Union hero turns against leadership," *Charleston Daily Mail,* May 30, 1990.

Nussel, Philip. "Health costs may cloud pension successes," *Charleston Daily Mail,* June 18, 1990.

Paynter, Teddy. "Phalen: Pittston won't be fair, equitable," *Logan Banner,* October 6, 1989.

Raffensperger, Gene. "Work begins on John L. Lewis Mining and Labor Museum," *Herald-Dispatch,* April 1, 1989.

Reed, David. "Labor Secretary Dole intervenes in strike; first face-to-face negotiations set today," *Williamson Daily News,* October 14, 1989.

Reed, David. "Judge refuses to waive UMW fines," *Herald-Dispatch,* February 13, 1990.

Rudolph, Barbara. "In the pits with Mr. Coal," *Forbes Magazine,* October 11, 1982, pp. 41-42.

Schechter Corporation v. United States, 295 U.S. 495 (1935).

Staff & Wire Reports. "New fines push union total past $30 million," *Bristol Herald Courier/Virginia-Tennessean,* October 6, 1989.

Staff Writer. "U. S. names W. J. Usery Jr. to mediate Pittston talks," *Wall Street Journal,* October 26, 1989.

Thurmond, Walter R. *The Logan Coal Field of West Virginia.* McClain Printing Company, West Virginia, 1964, p. 13.

Trumka, Richard L., International President, United Mine Workers of America. Memorandum addressed to all Executive Board Members, UMWA, in reference to District Conferences dated June 8, 1989.

Tyler, Patrick E. "National Bank of Washington: a Financial Citadel in Hot Water; Corporate Intrigue, Deception Documented," *Washington Post,* October 12, 1980, pp. A-1; A-10-11.

Tyler, Patrick E. "UMW's Power Play: The Union Takes Over the Bank's Board Room," *Washington Post,* October 12, 1980, p. A-11.

United Mine Workers Journal. June 15, 1969, pp. 3; 13.

United Press International. "Selective Strikes Begin," *Charleston Gazette,* August 22, 1989.

United Press International. "Judge again postpones UMW contempt trial," *Charleston Gazette,* December 9, 1989.

Watkins, Damon D. *Keeping the Home Fires Burning; A Book About The Coal Miner,* The Ohio Company, Ohio, 1937, pp. 91; 121.

Wells, Rob. "Miners mount horses to protest Pittston," *Williamson Daily News,* August 16, 1989.

Wells, Rob. "T-shirt art has strikers message," *Williamson Daily News,* August 21, 1989.

Wells, Rob. "UMW vote is set Monday. Both sides win concessions in new contract," *Williamson Daily News,* February 16, 1990.

Wessels, Andy. "Rockefeller fighting for UMW Benefits," *Charleston Gazette,* October 3, 1989.

Williams, Glen M., Judge, Order in the U. S. District Court for the Western District of Virginia, Abingdon Division, Civil Action #89-0090-A, September 20, 1989.